Biocompatibility Testing of Medical Devices

Biocompatibility Testing of Medical Devices

Shalinee Naidoo

ARCLER PRESS

www.arclerpress.com

Biocompatibility Testing of Medical Devices

Shalinee Naidoo

Arcler Press
224 Shoreacres Road
Burlington, ON L7L 2H2
Canada
www.arclerpress.com
Email: orders@arclereducation.com

ISBN: 978-1-77407-629-3 (Hardcover)

ABOUT THE AUTHOR

Shalinee Naidoo is currently the Regulatory and Product Development Manager of a medical device manufacturer based in South Africa. She is directly involved in global regulatory compliance and design and development of new medical devices from idea conception to market. She is also the founder of Life of Shal (www.lifeofshal.com) an online travel journal created to inspire others to explore the world and Scientist's Sanctuary (www.scientistssanctuary.com) – a science communication company that specializes in bridging the gap between scientific knowledge and creative communication for both the academic and corporate world.

TABLE OF CONTENTS

GLOSSARY

Absorb/absorption – The action of a foreign material or substance, or its decomposition products passing through or being assimilated by cells and/or tissue over time.

Acute systemic toxicity – Adverse effects occurring at any time after single, multiple or continuous exposures of a test sample within 24 hours.

Adverse Event – A problem that can or does result in permanent impairment, injury or death to the patient or the user.

Allergen – A substance or material that is capable of inducing a specific hypersensitive reaction.

Biomaterial – Any material or substance intended to interact with biological systems, it may be used to evaluate, treat, augment or replace any tissue, organ or function of the body.

Blood/device interaction – Any interaction between blood or any component of blood and a device resulting in effects on the blood, or on any organ or tissue, or on the device.

Body orifice – Any natural opening in the body. This includes the external surface of the eyeball or any permanent artificial opening.

Carcinogenic – Having the potential to cause cancer.

Certified reference material (CRM) – A reference material that serves as a control in experiments. These materials are usually accompanied by a certificate wherein one or more critical characteristics are certified by a procedure of validation testing through more than one laboratory.

Chronic systemic toxicity – Adverse effects occurring after the repeated or continuous administration of a test sample for a major part of the life span.

Clinical Data – Safety and/or performance information that are generated from the clinical use of a medical device.

Clinical Evaluation – The assessment and analysis of clinical data related to a medical device in order to verify the clinical safety and performance of the device when used as intended by the manufacturer.

Clinical Evidence – The clinical data and the clinical evaluation report pertaining to a medical device.

Clinical Evidence for an IVD Medical Device – All the information/data that supports the scientific validity and performance for its use as intended by the manufacturer.

Clinical Investigation – Any systematic investigation or study in or on one or more human subjects, undertaken to assess the safety and/or performance of a medical device. this term is similar in meaning to 'clinical trial' and 'clinical study.'

Coagulation – Phenomenon that results from activation of the clotting factor cascade.

Conformity Assessment Body (CAB) – A body other than a Regulatory Authority engaged in determining whether the relevant requirements in technical regulations or standards are fulfilled.

Degradation – Break down or decomposition of a material/substance.

Degradation product – any intermediate or final by-product which results from the physical, metabolic, and/or chemical decomposition of a material or substance.

Dose/dosage – An amount of test sample administered/expressed per unit of body weight or surface area

Dose-response – The relationship of dosage to the range of effects experienced.

Effective – The ability of a medical device or IVD medical device to provide clinically significant results in a significant portion of the target population.

Effectiveness – A device is clinically effective when it produces the effect intended by the manufacturer relative to the medical conditions. Active medical device: Any medical device whose operation depends on a source of electrical energy or any source of power other than that directly generated by the human body or gravity and which acts by converting this energy.

Erythema – Reddening of the skin or mucous membrane.

Eschar – Scab or discolored slough of skin.

Experimental control – A substance with well-characterized responses, which is used in a specific test system to assist in evaluating if the test system has responded in a reproducible and appropriate manner. The outcomes of such controls are already known.

Expiry Date/Expiration Date – Upper limit of the time interval during which the safety and performance characteristics of a material stored under specified conditions can be assured.

Extract – A liquid that results from extraction of the test sample or control.

Extract – Liquid or suspension that results from exposing a test or control material to a solvent under controlled conditions.

Extractables – A substance or substances that can be released from a medical device or material using extraction solvents and/or extraction conditions.

Genotoxicity – The process by which an agent/substance produces a deleterious effect on DNA and other cellular targets that control the integrity of genetic material.

Hematology – Study of the blood.

Harm – Physical injury or damage to the health of people or damage to property or the environment.

Hazard – Potential source of harm.

Hemocompatibility – The process used to assess the possibility of a test article to cause adverse effects on red blood cells (hemolysis), thrombosis, coagulation, platelets, and complement system.

Implantable device – Any device, including those that are partially or wholly absorbed and which is intended to be partially introduced into the human body through surgical intervention and intended to remain in place after the procedure for at least 30 days is also considered an implantable device.

***In Vitro* Diagnostic (IVD) Medical Device** – This is a medical device intended by the manufacturer for the in-vitro examination of specimens derived from the human body solely or principally to provide information for diagnostic, monitoring or compatibility purposes.

Incident – An unusual (unexpected) event associated with the use of a medical device that may or may not lead to problems.

Indications for Use – A general description of the disease or condition the medical device or IVD medical device will diagnose, treat, prevent, cure, or mitigate, including a description of the patient population for which the medical device or IVD medical device is intended.

Instructions for Use – Information provided by the manufacturer to inform the user of the medical device of the intended purpose and proper use and of any precautions to be taken. This is also known as a "package insert."

Intended use/purpose – The objective or intent of the manufacturer regarding the use of a product, process or service as reflected in the specifications, instructions, and information provided by the manufacturer.

Intracutaneous – Within the skin.

Intracutaneous testing – A test for immunity or allergy to a particular antigen by observing the local reaction following injection of a small amount of the antigen into the skin.

Invasive device – A device which, in whole or in part, penetrates inside the body, either through a body orifice or through the surface of the body.

Irritant – An agent that produces irritation.

Irritation – A localized non-specific inflammatory response to single, repeated or continuous application of a substance/material.

Label – Written, printed, or graphic information either appearing on the medical device itself, or on the packaging of each unit, or on the packaging of multiple devices.

Labeling – The label, instructions for use, and any other information that is related to identification, technical description, intended purpose and proper use of the medical device, but excluding shipping documents.

Lay User – An individual who does not have formal training in a relevant field or discipline.

Leachable substance – A chemical removed from a device or material by the action of water or other liquids, usually occurring during use of the device.

Long term – Normally intended for continuous use for more than 30 days.

Manufacturer – Any natural or legal person with the responsibility for design and/or manufacture of a medical device with the intention of making the medical device available for use, under their name; whether or not such a medical device is designed and/or manufactured by that person themselves or on their behalf by another person(s).

Medical Device – Any instrument, apparatus, machine, appliance, implant, reagent for *in vitro* use, software, material or other similar or related article, intended by the manufacturer to be used, alone or in combination, for human beings, for one or more of the specific medical purpose(s) such as: diagnosis, prevention, monitoring, treatment or alleviation of disease, etc. This device does not achieve its primary intended action by pharmacological, immunological, or metabolic means, in or on the human body, but which may be assisted in its intended function by such means.

Necrosis – Cell death that occurs as a direct result of irreversible changes caused by injury or disease.

negative control – Any well-characterized material and/or substance, which, when tested by a specific procedure, demonstrates the suitability of the procedure to yield a reproducible, appropriately negative, non-reactive or minimal response in the test system.

Negative control – Any well-characterized material or substance that, when tested by a specific procedure, demonstrates the suitability of the procedure to yield a reproducible, appropriately negative, non-reactive or minimal response in the test system.

Edema – Swelling due to abnormal infiltration of fluid into the tissues.

Patient – An individual under the care of a healthcare provider who may benefit from the action of a medical device.

Performance – The ability of a medical device to achieve its intended purpose as stated by the manufacturer.

Pharmacokinetics – The branch of pharmacology concerned with the movement of drugs within the body

Platelet – Cellular body that is present in the circulation which adheres to surfaces and aggregates to form a hemostatic plug which minimizes bleeding.

positive control – Any well-characterized material and/or substance, which, when evaluated by a specific test method, demonstrates the suitability of the test system to yield a reproducible, appropriately positive or reactive response in the test system.

Positive control – Any well-characterized material or substance that, when evaluated by a specific test method, demonstrates the suitability of the test system to yield a reproducible, appropriately positive or reactive response in the test system.

Post market – When the products are already on the market.

Reference material – A material with one or more properties whose values are sufficiently reproducible and well established. This enables use of the material or substance for the calibration of an apparatus, the assessment of a measurement method, or for the assignment of values to materials. When tested by the test procedure it demonstrates the suitability of the procedure to yield a reproducible, predictable response which may be negative or positive.

Regulatory Authority (RA) – A government body or other entity that exercises a legal right to control the use or sale of medical devices within its jurisdiction, and that may take action to ensure that medical products marketed within its jurisdiction comply with legal requirements.

Risk – Combination of the probability of occurrence of harm and the severity of that harm.

Risk Analysis – Systematic use of available information to identify hazards and to estimate the risk.

Risk Assessment – Overall process comprising a risk analysis and a risk evaluation.

Risk Evaluation – Procedure based on the risk analysis to determine whether tolerable risk has been exceeded.

Safety – Free from unacceptable risk.

Shelf-Life – The period of time until the expiry date during which a medical device in its original packaging maintains its stability under the storage conditions specified by the manufacturer.

Short term – Normally intended for continuous use for not more than 30 days.

Skin sensitization – An immunologically mediated cutaneous reaction to a substance.

Stability – The ability of a medical device and IVD medical device to maintain its safety and performance characteristics within the manufacturer's specifications over a specified period of time.

Subchronic systemic toxicity – Adverse effects occurring after the repeated or continuous administration of a test sample for a part of the lifespan of test animal.

Subacute systemic toxicity – Adverse effects occurring after multiple or continuous exposure between 24 hours and 28 days.

Surgically invasive device – An invasive device which penetrates inside the body through the surface of the body, with the aid or in the context of a surgical operation.

Systemic toxicity – Toxicity that is not limited to adverse effects at the site of contact between the body and the device. Toxicity occurs as a result of absorption and distribution of a toxic substance from its entry point to a distant site at which deleterious effects are produced.

Test sample – A material, device, device portion, component, extract or portion thereof that is subjected to the biological or chemical testing or evaluation.

Test sample – A medical device, component or material that may serve a representative of a medical device, or an extract or portion thereof that is subjected to the biological or chemical testing or evaluation.

Thrombosis – Partial or complete occlusion of a vessel or device by a thrombus.

Transient – Normally intended for continuous use for less than 60 minutes.

Ulceration – open sore representing loss of superficial tissue.

Vehicle – A liquid used to moisten, dilute, suspend, extract or dissolve the test substance/ material.

LIST OF FIGURES

LIST OF TABLES

LIST OF BOXES

LIST OF ABBREVIATIONS

AAS	Atomic Absorption Spectroscopy
AB	Alamar Blue
ADME	Absorption/Distribution/Metabolism/Excretion
AFC	Amino Fluoro Coumarin
ANSI	American National Standards Institute
ASTM	American standards of testing and materials
ATP	Adenosine Triphosphate Assay
BAI	Biomaterial Associated Infections
BEP	Biological Evaluation Plan
CBC	Complete Blood Count
CF	Carboxyfluorescein
CFA	Complete Freund's Adjuvant
CFDA-AM assay	5-carboxyfluorescein diacetate, acetoxymethyl ester assay
CO_2	Carbon Dioxide
CRM	Certified Reference Materials
CVS assay	Crystal Violet Assay
DMSO	Dimethyl Sulfoxide
ECH	Ethylene Chlorohydrin
EDX	Energy-Dispersive X-Ray Analysis
EP	Essential Principles
EtO	Ethylene Oxide
FDA	Food and Drug Administration
GF-AFC	Glycylphenylalanyl-Aminofluorocoumarin
GLP	Good Laboratory Practice
GPMT	Guinea Pig Maximization Test
ICP	Inductively-Coupled Plasma Spectroscopy
IDE	Investigational Device Exemption

IEC	International Electrotechnical Commission
IR	Infrared Spectroscopy
ISO	International Standards Organization
ITU	International Telecommunication Union
IVD	*in vitro* Diagnostic
LAL	Limulus Amoebocyte Lysate
LDH	Lactate Dehydrogenase
LLNA	Local Lymph Node Assay
MEM	Minimal Essential Medium
MHRA	Medicines and Healthcare products Regulatory Agency
MK	Magnusson and Kligman
MRI	Magnetic Resonance Imaging
MTS assay	(5-(3-carboxymethyoxyphenyl)-2-(4-5-dimethyl-thiazoly)-3-(4-sulfophenyl) tetrazolium inner salt assay
MTT assay	3-(4,5-dimethylthiazol-2-yl)-2-5-diphenyltetrazolium bromide) assay
NRU assay	Neutral Red Update Assay
OD	Optical Density
OECD	Organization for Economic Co-operation and Development
OOS	Out of Specification
PEG	Polyethylene Glycol
PII	Primary Irritation Index
PMA	Premarket Approval
PT	Prothrombin Time
PTT	Partial Thromboplastin Time
QAU	Quality Assurance Unit
RBC	Red Blood Cells
RM	Reference Materials
SCC	Standards Council of Canada
SEM	Scanning Electron Microscopy
SME	Small to Medical Enterprise
SOP	Standard Operating Procedure
SPU-ZDBC	Segmented Polyurethane Rod Dibutyldithiocarbamate
SRB assay	Sulforhodamine B Assay
TEM	Transmission Electron Microscopy

TER	Transcutaneous Electrical Resistance
TFT	Triflurothymidine
UDS	Unscheduled DNA Synthesis
USP	United States pharmacopeia
WBC	White Blood Cells
WST-1 assay	2-(4-iodophenyl)-3-(4-nitrophenyl)-5-(2,4-disulfophenyl)-SH tetrazolium monosodium salt) cell proliferation assay
WST-8 assay	2-(2-methoxy-4-nitrophenyl)-3-(4-nitrophenyl)-5-(2,4-disulfophenyl)-2H tetrazolium, monosodium salt) assay
XTT assay	(2,3-bis(2-methoxy-4-nitro-5-sulphophenyl)-5-carboxanilide-2H-tetazolium, monosodium salt

PREFACE

Biocompatibility is, by definition, a measurement of how compatible a device is with a biological system. Ensuring that a finished device is safe for human use and does not result in any harmful effects is imperative before being placed on the market. Manufacturers must be able to show the safety of all device components and the finished device itself in the form of a master file holding acceptable toxicological data and related literature. If a manufacturer cannot show this information, then their device will need to undergo testing. To help manufacturers understand biocompatibility testing requirements, a range of standards have been published with the goal of establishing the minimum attributes necessary to ensure international safety. This volume aims to provide a simple understanding around the concepts of biocompatible medical devices. Much information is provided to help manufacturers when choosing appropriate tests and to ensure that biomaterials and finished devices are safe and will perform as intended when used in a clinical setting. Key concepts of the ISO 10993 series of standards are provided as well as some of the major challenges faced by medical device manufacturers when considering testing. A simple understanding around the evaluation and characterization of biocompatible materials as well as the regulation of such devices are provided. In addition, an overview of how to set up a basic biocompatibility testing laboratory is also included.

CHAPTER 1

Understanding Medical Devices

CONTENTS

1.1. INTRODUCTION

Medical devices are essential in everyday life. From diagnostics to the prevention and treatment of diseases to contraception, medical devices play a defining role. Some have close contact with tissues or blood and may be implanted in the body while others function as surface or external communicating devices. For this reason, it is imperative to ensure that these devices have no detrimental effects when in contact with humans. The types of safety or effectiveness tests used for medical devices have, in the past, varied among national regulatory agencies across different territories. Prior to the 1990s, individual countries were involved in developing their own guidelines which specified the type of toxicological studies essential for registration of medical devices in that region. The United States, Canada, and Great Britain had also collaborated in the issuance of a tripartite agreement of testing requirements among the three nations. Various national standards groups such as the American Standards of Testing and Materials (ASTM 1995), Canadian Standards Association, and British Standards Institute had also developed guidance documents and standards on individual assays for testing of medical devices. In the late 1980's the process of developing international standards for medical devices began. This was under the guidance of the International Standards Organization (ISO) whose main goal was to harmonize the requirements among participating nations so as to support international trade. This led to the development of a series of 20 standards focusing on the preclinical and clinical evaluation of medical devices. These standards, better known as the ISO standards, form a set of consensus standards whose minimum requirements are agreed upon by all participating nations. These form the basis for testing, in addition to any individual country-specific requirements.

Biocompatibility is by definition, a measurement of how compatible a device is with a biological system. The purpose of performing biocompatibility testing is to determine the safety of a device for human use, taking into consideration the intended use and specific legal requirements at the time of registration. This is essential to determine whether a device has the potential to cause adverse effects as many devices may be subject to degradation when implanted into tissues or subjected to the surrounding degradation effects of body. For a manufacturer, it is important to note that biocompatibility testing will be required unless there is sufficient history of prior use of the device when on the market, proving safety of all device components and the finished device itself. Manufacturers must then be able to show a master file with acceptable toxicological data and literature. If a manufacturer cannot show this information, then their device will need to undergo testing. Determining the biocompatibility testing requirements for a medical device involves multiple levels of decisions and a thorough understanding of the functionality of the device itself. The site and duration of body contact, prior use of the material in a medical device, the type of material(s) used, and available information on the toxicological profile of the materials must all be defined and available. Testing requirements are guided by the ISO 10993 series of standards. These standards have been developed for assessing cytotoxicity, sensitization, irritation, systemic toxicity, subchronic toxicity, genotoxicity, implantation, hemocompatibility, chronic toxicity, carcinogenicity, reproductive toxicity, and degradation of devices. Specific methodology is provided for each assay as well as cross-referencing to other existing standards where applicable. Manufacturers should source a suitably accredited laboratory skilled in biocompatibility testing of medical devices. These laboratories should not only be able to test medical devices but should also be able to provide support and guidance on which biocompatibility tests are recommended for the device in question.

1.2. WHAT IS A MEDICAL DEVICE?

A medical device is defined as a product or equipment that is intended for use within the health care industry to either diagnose, monitor or treat diseases or to act as a supportive aid for people with any form of illness or disability. There are over 500,000 different types of medical devices and these include a vast array of devices such as x-ray machines, infusion pumps and cardiac pacemakers used by doctors, nurses, and surgeons as well as common devices that may be found every day in homes such as wound dressings, thermometers, contact lenses and wound irrigation solutions. According to ISO13485, medical devices range from simple low-risk products such as urine bags and latex gloves to more complex, technologically advanced programmable devices such as pacemakers. Generally, these devices are grouped into four classes namely class I, class IIa, class IIb and class III; with class I being the lowest risk and class III being the highest risk (Figure 1.1) or Class A, B, C, D with Class A being the lowest risk and Class D being the highest risk.

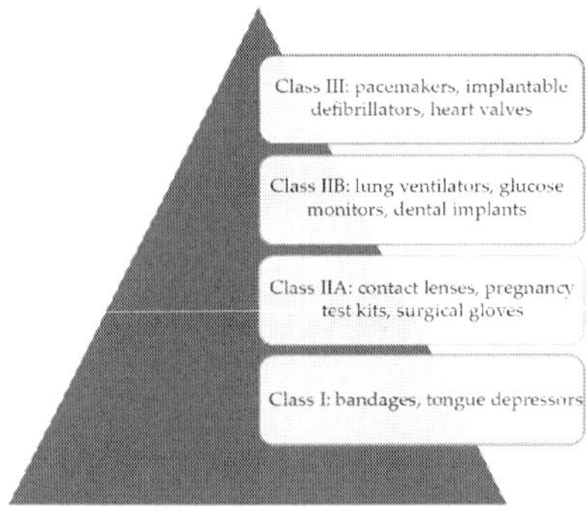

Figure 1.1. Overview of the different medical device classes.

The World Health Organization defines a medical device as any instrument, apparatus, implement, machine, appliance, implant, reagent for *in vitro* use, software, material or other similarly related article intended by the manufacturer to be used, alone or in combination and for human beings in one or more specific medical purpose(s) (Table 1.1). These devices should not achieve its main intended function by pharmacological, immunological, or metabolic means on or in the human body. It may however be assisted in its intended function by the abovementioned methods.

Table 1.1. General Uses for Medical Devices

- Diagnosis, prevention, monitoring, treatment or alleviation of disease;
- Diagnosis, monitoring, treatment, alleviation of, or compensation for an injury;
- Investigation, replacement, modification, or support of the anatomy and/or of a physiological process;
- Supporting or sustaining life;
- Control of conception;
- Disinfection of medical devices;
- Providing information by means of *in vitro* examination of specimens derived from the human body; and
- Does not achieve its primary intended action by pharmacological, immunological or metabolic means, in or on the human body, but maybe assisted in its intended function by such means.

Source: Derived from WHO (https://www.who.int/medical_devices/full_deffinition/en/).

The mode of action of a medical device on the human body is what differentiates it from medicines, whose mode of action may often be metabolic, immunological, or pharmacological. In fact, definitions must make clear distinctions on the difference between a medical device and medicine/drug whose primary intended use is obtained through chemical action or by being metabolized in the body. While many regulatory bodies tend to establish their own definitions around this topic, the Global Harmonization Task Force has proposed a harmonized definition for medical devices and in-vitro medical devices to ensure uniformity of understanding on a global scale (GHTF document SG1/N029R11).

Box 1: Definition of a Medical Device as Defined by GHTF (GHTF/SG1/N071:2012)

"Medical device" means any instrument, apparatus, implement, machine, appliance, implant, *in vitro* reagent or calibrator, software, material or other similar or related article, intended by the manufacturer to be used, alone or in combination, for human beings for one or more of the specific purposes of:

- Diagnosis, prevention, monitoring, treatment or alleviation of disease;
- Diagnosis, monitoring, treatment, alleviation of or compensation for an injury;
- Investigation, replacement, modification, or support of the anatomy or of a physiological process;
- Supporting or sustaining life;
- Control of conception;
- Disinfection of medical devices;
- Providing information for medical purposes by means of *in vitro* examination of specimens derived from the human body; and
- Which does not achieve its primary intended action in or on the human body by pharmacological, immunological, or metabolic means, but which may be assisted in its function by such means.

The definition laid out by the GHTF covers a multitude of different medical devices, some of which are complex and reflect the latest advances in technological developments such as imaging equipment and implants; as well as those simple devices such as tongue depressors, thermometers, scales, and latex gloves. The GHTF document further defines those produces which may be considered to be medical devices in some jurisdictions while not classified in others. These include (1) disinfection substances, (2) aids for persons with disabilities, (3) devices incorporating animal and/or human tissues, and (4) devices for in-vitro fertilization or assisted reproduction technologies.

Box 2: Definition of an *In-Vitro* Medical Device as Defined by GHTF (GHTF/SG1/N071:2012)

> '*In Vitro*' Diagnostic (IVD) medical device' means a medical device, whether used alone or in combination, intended by the manufacturer for the in-vitro examination of specimens derived from the human body solely or principally to provide information for diagnostic, monitoring or compatibility purposes.

1.3. UNDERSTANDING THE HISTORY OF MEDICAL DEVICES

Medical devices have been around for centuries. In fact, evidence dates far back as 7000 BC to the use of scalpels, slings, and splints, which, in essence, are medical devices used in health care. As modern technology and the application of medical devices merged, so their applications took off in the last fifty years; thereby making the medical devices an essential part of health care and patient safety. The key landmarks and trends in medical device development are highlighted (Figure 1.2).

1800 - 1850
- FIRST MODERN STETHOSCOPES, LARYNGOSCOPES, OPTHALMOSCOPES

1895
- X-RAYS: discovered by German physicist Wilhelm Roentgen

1903
- FIRST ELECTROCARDIOGRAPH: developed by Dutch physician and physiologist Willem Einthoven. He was awarded the Nobel Prize in 1924 for his discovery

1927
- FIRST MODERN RESPIRATOR: Devised by the United States medical researcher Philip Drinker and his collegues at Harvard University

1928
- FIRST CARDIAC CATHETERIZATION: performed by Werner Forssmann on himself to show feasibility of technique for injecting drugs directly into the heart

1945
- 1940: FIRST METALLIC HIP REPLACEMENT SURGERY: performed by United States surgeon Dr. Austin T. Moore
- 1945: FIRST KIDNEY DIALYSIS MACHINE: invented by Dutch-born physician Willem Kolff

1950
- FIRST ARTIFICIAL HIP REPLACEMENT (ARTHROPLASTY): Performed by English surgeon Sir John Charnley

1951 — •FIRST COMMERCIALLY AVAILABLE ARTIFICIAL HEART VALVE: invented by a United States team led by electrical engineer Miles Edward

1952 — •FIRST SUCCESSFUL (EXTRENAL) CARDIAC PACEMAKER: Developed by United States cardiologist Paul Zoll

1960 — •FIRST TOTALLY INTERNAL PACEMAKER: Developed by United States electrical engineer Wilson Greatbatch

1970 — •FIRST CIMPUTERIZED AXIAL TOMOGRAPHY (CT) SCANNER: Dveveloped by British electrical engineer Godfrey Hounsfield and South Africa born physicist Allen Cormack (jointly awarded the Nobel Prize for medicine in 1979)

1976 — •FIRST REGULATORY SYSTEM FOR MEDICAL DEVICES: created by the United States government

1976 — •FIRST POSITRON EMISSION TOMOGRAPHY (PET) SCANNING OF PEOPLE: Performed by Abass Alavi, University of Pennsylvania

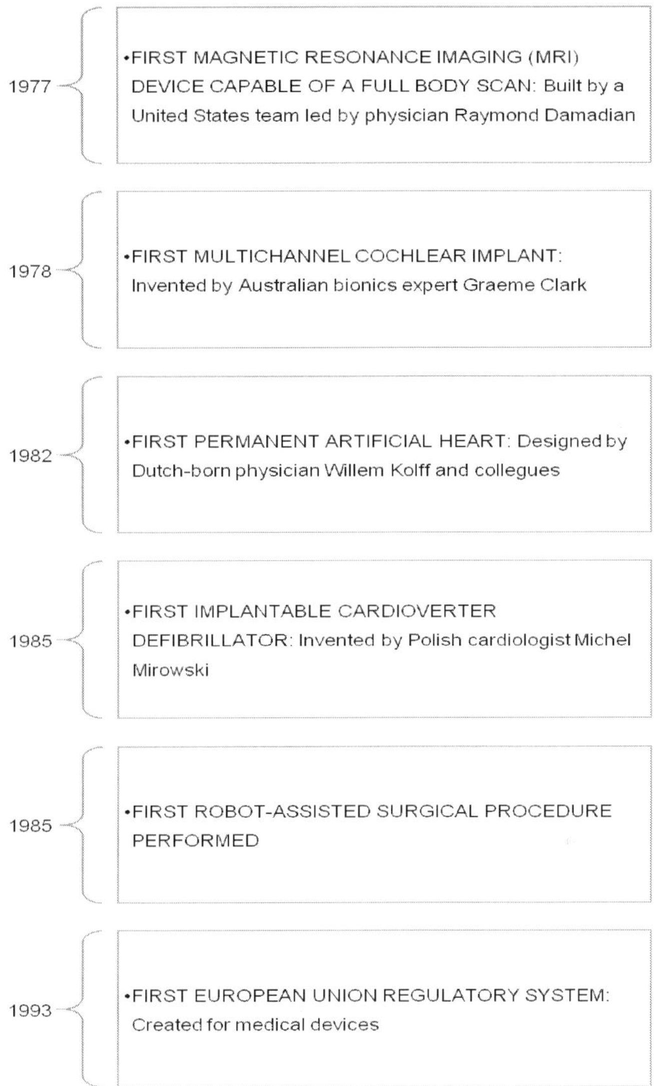

Figure 1.2. Overview of landmarks and key trends in medical device development.

The 1980s saw a surge in the number of patient care medical devices with particular focus on high-resolution imaging devices. In addition, systems for continuous monitoring of cardiovascular parameters such as heart rate and blood pressure were becoming increasingly more standardized in hospitals and health care facilities. In addition, many health care treatments were being taken over by technological advances in medical devices such as ventilators, kidney dialysis machines, and neonatal incubators. The 1980s to 2000s saw more hospitals in industrialized countries adopting CT scanners and magnetic resonance imaging (MRI) units. The choice of medical devices that could be used grew drastically. Between the 2000s and 2010, robotics became a reality within the medical device industry. Assistive devices that could be used with people that had functional disabilities increased. In addition, the use of medical device integrated with information systems or web-based systems became more and more popular.

1.4. THE INFLUENCE OF TECHNOLOGY TRENDS ON MEDICAL DEVICE DEVELOPMENT

As technology advances so does the development of smaller and less expensive robotic systems that ideally allow high-precision surgeries to continue. The need for synergy will continue to direct future innovation and medical device design while nanotechnology and genomics will interact and continue to improve personalized care. Research and development officers within manufacturing firms may be swayed by the lure of technology however ideally what is needed is a balance between technology, cost-effectiveness, need, and overall usefulness of the device in industry. One of the biggest examples supporting this trend is among the medical imaging facilities whose main aim is to promote whole-body CT scanning as a means for preventative measures among individuals with no symptoms or suspected disease implications. The US FDA announced in July 2009, that it had not cleared or approved any CT system specifically for use in screening as no manufacturer has ever been able to demonstrate effective disease screening with their CT scanner. In addition, they also made reference to the risk of radiation doses through repetitive CT examinations. As technology changes and improves, so trends develop; all of which have hidden implications on the use of the medical devices and their expected health implications.

Table 1.2. Current Trends Which Have Had or Will Continue To Have a Significant Effect on Medical Device Use and Related Implications for Health

Technology convergence	Medical device implications	• Labor-intensive; • Costly; • Requires multi-disciplinary teams; • Highly dependent on infrastructure; • Difficult to apply in low-resource settings; • Likely to raise patient expectations; • Patient pressure on health care; • Likely to be reserved for financially privileged patients; • Liable to incur technical risks; and • Patient safety problems may arise.
	Health implications	• Could enhance sharing of clinical information; • Could reduce errors of data entry; • Could facilitate data analysis; • Could enhance clinician efficiency; • Could allow increased case load; • Could improve patient outcomes; and • Could improve patient safety by giving clinicians access to all patient data.
Decentralization of care delivery	Medical device implications	• Reliance on portable devices. • May result in requiring of greater ruggedness than stationary hospital equipment; and • Reliance on non-medical technology such as communication networks, durable battery technology, and power sources
	Health implications	• Benefit to patients and family caregivers; • Adequate user training required; • May result in more patients receiving care outside a hospital setting; • More efficient patient monitoring; • More efficient health care; and • Enables access of rural health-care professionals to highly specialized clinical procedures
Computer-aided surgery and robotics	Medical device implications	• Risk of use error; • Risk of system malfunctioning; • May lead to patient injury; • Longer learning curve for users; • Requires integration of imaging and surgical navigation systems; • Device-specific training required; • High acquisition costs; • High procedure costs; • Could be prohibitive for many resource-scarce settings; and • Maintenance and quality control requirements likely to be demanding.

	Health implications	• Greater accuracy of surgical procedures; • Long initial learning curve; • Could still save time; • Greater consistency where reproduction of procedures is concerned; • Could circumvent obstacles to surgical intervention; • Allows frequent checking of instrument status and positioning; and • Could give nurses more time with patients by alleviating menial tasks

1.5. THE GLOBAL MEDICAL DEVICE MARKET

The global medical device market is enormous. In 2008, revenue from sales of medical devices worldwide was estimated at little over US$ 210 billion. This is close to double that of the estimated revenue for 2001. Globally, the medical device industry comprises of more than 27,000 medical device companies. Four fifths of the global medical device sales revenue comes from sales in the United States and Europe with around 10 countries accounting for approximately 80% of global sales revenue (Figure 1.3).

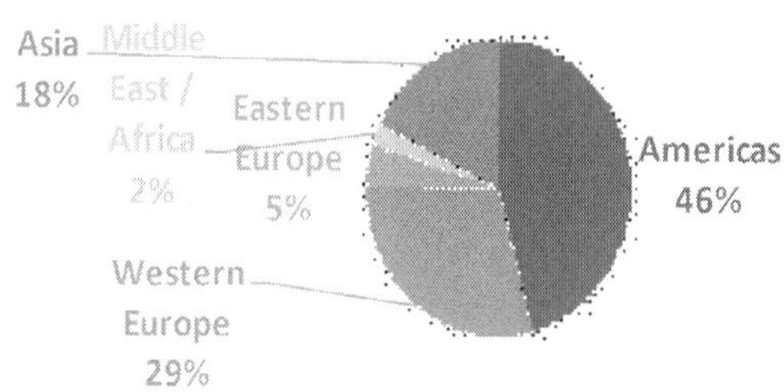

Figure 1.3. Medical device markets by region based on 2009% sales revenue (Ref.: The World Medical Markets Fact Book, 2009).

The United States is at the top of that list accounting for around 40% of sales, closely followed by Japan (10%), Germany (8%) and France (4%).

Table 1.3. Top Ten Countries by 2009 Sales Revenue

Ranking	Country	Sales Revenue US$ (Millions)	Percent (%)
1	United States	91,316	40.7%
2	Japan	22,721	10.1%
3	Germany	18,147	8.1%

4	France	8,625	3.8%
5	Italy	8,004	3.8%
6	United Kingdom	7,628	3.4%
7	China	6,161	2.7%
8	Spain	4,887	2.2%
9	Canada	4,757	2.1%
10	Switzerland	4,063	1.8%
Subtotal		176,309	78.6%
World Total (67 Countries)		**224,103**	**100%**

Source: The World Medical Markets Fact Book (2009).

There are about 27,000 medical device companies in the world. Sales revenues indicate that the majority of the top 30 medical device companies have their headquarters in the United States of America (Table 1.3). Together, these companies account for 89 and of the estimated global sales revenue of US$ 210 billion. The remaining 11% is shared by a vast number of manufacturers in the small to medical enterprise (SME) category.

Table 1.4. Top 30 Medical Device Companies Based on Sales Revenue

Ranking	Company	Headquarters	Sales Revenue US$ (Millions)
1	Johnson & Johnson	United States	23,225
2	GE Healthcare	United States	17,392
3	Siemens Healthcare	Germany	15,526
4	Medtronic	United States	13,515
5	Baxter International	United States	12,400
6	Covidien	Ireland	9,910
7	Philips Healthcare	Netherlands	9,227
8	Boston Scientific	United States	8,050
9	Becton Dickinson	United States	7,156
10	Stryker	United States	6,718
11	B. Braun	Germany	5,263
12	Cardinal Health	Ireland	4,600
13	St. Jude Medical	United States	4,363
14	3M Health Care	United States	4,293
15	Zimmer	United States	4,121

16	Olympus	Japan	3,920
17	Smith & Nephew	United Kingdom	3,801
18	Hospira	United States	3,620
19	Terumo	Japan	3,400
20	Danaher Corporation	United States	3,227
21	Synthes	United States	3,206
22	Beckman Coulter	United States	3,099
23	Alcon	Switzerland	2,881
24	Fresenius Medical Care	Germany	2,875
25	CR. Bard	United States	2,452
26	Abbott	United States	2,241
27	Dentsply	United States	2,194
28	Varian Medical	United States	2,070
29	Biomet	United States	2,135
30	Drager	Germany	1,729
			185,734

Source: Derived from company annual reports (listed in the references section). It should be noted that not all data refers strictly to medical devices. For those companies that compete in several industrial sectors, there is no differentiation between sales revenues from medical devices and other products.

Historically, however, the majority of high-tech medical devices have been manufactured by companies based in first world/industrialized countries while low tech devices (such as condoms, surgical gloves, gauze, and so on) have been manufactured in countries with emerging economies such as India, Indonesia, Malaysia, and Sri Lanka. In addition, many of the multinational companies listed in Table 1.4 tends to have manufacturing sites based in developing countries due to the cheaper costs (Table 1.5).

Table 1.5. Top Medical Device Manufacturers by Country Based on Sales Revenue

Ranking	Country	Sales Revenue US$ (Millions)	Percent (%)
1	China	6,161	28.6%
2	Brazil	2,606	12.1%
3	Mexico	1,890	8.8%
4	India	1,617	7.5%
5	Turkey	1,062	4.9%

6	Malaysia	826	3.8%
7	South Africa	701	3.2%
8	Thailand	661	3.1%
9	Colombia	530	2.5%
10	Iran	465	2.2%
11	Argentina	419	1.9%
12	Egypt	416	1.9%
13	Venezuela	371	1.7%
14	Romania	355	1.6%
15	Cuba	345	1.6%
16	Chile	309	1.4%
17	Vietnam	288	1.3%
18	Croatia	255	1.2%
19	Belarus	253	1.2%
20	Ukraine	249	1.1%
21	Bulgaria	229	1.1%
22	Lithuania	201	0.9%
23	Serbia	199	0.9%
24	Indonesia	194	0.9%
25	Pakistan	184	0.8%
26	Peru	183	0.8%
27	Philippines	163	0.8%
28	Morocco	152	0.7%
29	Jordan	144	0.6%
30	Latvia	141	100%
Subtotal		21,569	100%
World Total (67 Countries)		**224,103**	**100%**

Source: Based on The World Medical Markets Fact Book (2009).

1.6. UNDERSTANDING THE DIFFERENCES BETWEEN MEDICINES AND MEDICAL DEVICES

Medicines and medical devices are actually quite similar in various aspects. Both form health technologies which society needs to survive and ensure healthy living, both can be used to diagnose, treat, alleviate, and cure diseases. Both require some form of regulatory control and both need an adequately implemented post-market surveillance system. In addition, both have some form of intellectual property, need a supply chain, and have become crucial to modern health care. Despite the numerous similarities, medical devices differ from medicines in various ways.

Table 1.6. Key Points in the Differences Between Medical Devices and Medicines

Diversity	**Medical Devices Vary in Size, Complexity, Packaging, and Use**
Innovation	Innovation of medical devices results mainly from the insights of clinicians rather than laboratory studies. Medical devices usually undergo small/incremental improvements. They have a relatively short commercial lifecycle of about 18 months on average.
Durability	Medical devices have a wide range of durability with extremes ranging from a few minutes for disposable devices, to several decades for some implantable devices and medical equipment.
Mode of action	Medical devices do not achieve their principal intended action in or on the human body by pharmacological, immunological, or metabolic means, although some devices such as syringes may be used to deliver medicines. Medical devices produce mainly local and physical effects on the body rather than systemic and pharmacological effects.
Regulation	The risk class of the medical device in question will determine the extent of regulatory scrutiny involved. Often the assessment of safety and efficacy for low-risk classes of medical devices can be performed by the manufacturer themselves. For high-risk classes of medical devices, some form of evidence may be submitted to the competent authorities to prove safety and efficacy. Efficacy or effectiveness of medical devices is proven before they are put on the market, but it should be noted that clinical effectiveness is more difficult to prove.
Supply	About 80% of the medical device industry is made up of small and medium enterprises. Distribution of heavy medical equipment is usually costly. There is no well-defined supply chain or profession involved in the supply of medical devices which is direct contrast to that of medicines where pharmacists may aid in the supply.

Usage	The performance of a medical device depends not only on the device itself but also on how it is used. The user interface of a medical device is usually not direct; for example, device-to-patient interaction. In many cases, there is an intermediary interface involving a device-operator-patient interaction. Often medical devices are associated with some form of learning. This is particularly true with more complex high-tech devices which may need a form of technical training and support. Medical devices may require service and maintenance. Many medical devices are used for diagnostic purposes. Many medical devices are used to alleviate functional disabilities, and this must be taken into consideration.

Medical devices are often based on mechanical, biomedical engineering techniques and may either, permanently or temporarily replace a body function while medicines on the other hand are designed to interact with the body's chemistry. The efficacy of medicine, on the one hand, is much simpler to demonstrate than for a medical device as this depends on many more variables such as the setting in which the device will operate, the individual patient characteristics as well as the skill and experience of the health care practitioner using or applying the device. Furthermore, the research and development models of medical devices as compared to medicines are extremely different. Most medical devices cannot be evaluated using a randomized clinical trial as is often the case for medicines. They are, however, evaluated for their overall efficacy and safety based on data acquired from clinical investigations.

While both, medical devices and medicines are regulated, the medical device industry has its own unique regulatory system when compared to medicines. Medical devices tend to have shorter product life cycles with technical improvements taking an estimate of two years as opposed to decades for medicines. For this reason, in particular, trying to apply the same regulations of medicines to medical devices would thus delay access to vital medical resources and procedures while not really increasing patient safety. Lastly, it should be noted that the sales and marketing models differ for medical devices as when compared to medicines. Where medicines are concerned, it is mainly the prescribing doctor who becomes the target of sales and marketing; on the other hand, for medical devices; there are many more stakeholders concerned who can influence the overall adoption and use of a device. These often include hospital managers, nurses, and so on.

For this reason, in particular, medical devices have a greater emphasis on training, education, service, and overall distribution and maintenance; all of which can greatly affect business models (Table 1.7)

Table 1.7. Detailed Differences between Medical Devices and Medicines

Medical Devices	***In Vitro* Diagnostic Medical Devices**	**Medicines**
In vivo and/or *ex vivo* uses	*In vitro* use	*In vivo* use
Diagnostic or therapeutic intended use	Diagnostic intended use	Therapeutic intended use
Outcomes of use often depend directly on skill or experience of user	Outcomes generally not dependent on skill or experience of user	Outcomes generally not dependent on skill or experience of user
Active components: Generally based on mechanical, electrical, and materials engineering. Many medical devices incorporate and are driven by software.	Active components: IVD components have no therapeutic effect. They are used only for diagnosis. The key components are those that are essential for detection of the analyte of interest. Core reagents are biological (for example, antibodies) Performance of various tests (such as sensitivity and/or specificity) depends on the design of test, geographic variations of the infective agent, populations, and the setting of use. Variable batch sizes for a given reagent. Individual batches of the same reagent may use different starting materials. Stability varies between products and may vary between batches. Generally stored at 4°–8°C Generally short shelf lives (< 12 months)	Active components: Based on pharmacology and chemistry; now encompassing biotechnology, genetic engineering, and so on. Pharmacologic properties and action of active ingredients are known. These are based on pre-clinical and clinical studies. Standardized batch sizes, manufacturing processes and starting materials. Products are stable. Products are generally stored at room temperature. Generally, have long shelf lives.

Product development: Can be used for a wide variety of products and applications from thermometers and bandages to pacemakers to x-rays. Designed to perform specific functions and approved on the basis of safety and performance. Many products developed by doctors or nurse	Product development: Products are usually in the form of reagents. Wide variety of IVDs designed for different indications such as screening, confirmation, and monitoring of treatment. Product development by discovery and evaluation. Approved on basis of performance. Products developed in laboratories by chemists and biologists.	Product development: Products are usually in the form of pills, solutions, aerosols, or ointments. Product development by discovery and trial. Approved on basis of safety and efficacy. Products developed in laboratories by chemists and pharmacologists.
Most act through physical interaction with the body or body part.	Tests performed on samples (such as blood, tissues, saliva, feces or urine) that have been taken from the body. No direct contact with human body. No need for clinical trials with patients. Generally, "performance evaluations" of IVDs conducted with samples.	Products are administered by mouth, skin, eyes, inhalation, or injection and are biologically active. Products are effective when absorbed into the human body. Products often act systemically on the entire body
Concerns around intellectual property: Continuous innovation and iterative improvements based on new science, new technology, and new materials.	Concerns around intellectual property: Continuous innovation and iterative improvements based on new science, new technology, and new materials.	Concerns around intellectual property: Extensive research and development of a specific compound or molecule; takes several years for a new drug to enter the product pipeline.
Short product life cycle and investment recovery period (typically 18 months on market). Little patent linkage possible. Data exclusivity is important.	Rapid innovation, short product life cycle (3–5 years) and investment recovery period.	Intensive patent protection, including data exclusivity and patent linkage, needed due to extensive product life cycle and long investment recovery period.

Majority of new products bring added functions and clinical value based on incremental improvements; this often results in a range of related "models" for a user to choose among based on patient needs.	Majority of new products bring added clinical value based on improvements of performance	Usually large step innovation.
Support provided: Large investment in manufacturing, distribution, and training/education (and retraining). Ned to provide service and maintenance (for many high-tech devices).	Support provided: Large investment in manufacturing, distribution, and training/education (and retraining). Need to provide service and maintenance (for many high-tech devices).	Support provided: Low manufacturing and distribution cost, and, in most cases, little or no training, service, or maintenance costs.

1.7. IMPORTANT FACTORS TO CONSIDER WHEN CHOOSING A MEDICAL DEVICE

The process of choosing a medical device is a complex one and often requires a transparent process that is based on information, reasoning, evidence, assessment of public health needs, and a process of priority. The need to improve individual and public health must always be taken into consideration. If this is not done, it can lead to inappropriate use and an overall waste of financial resources. The abovementioned factors tend to have negative consequences in both industrialized and developing countries. One such example is diagnostic imaging. The WHO estimates that high-tech diagnostic imaging is required only in 20–30% of medical cases worldwide. Of those cases, around 80–90% of diagnostic problems can generally be solved using basic x-ray and/or ultrasound examinations.

With the multitude of manufacturers available worldwide, each of whom produces various kinds of medical devices on a daily basis; it is understandable that choosing an appropriate device that is safe and which performs as intended has become difficult for ministries of health, regulatory authorities, managers, and health care facilities. In addition, the lack of professional technical advice from medical device specialists also contributes to these difficulties. There are various barriers to choosing medical devices such as fascination with technology, aggressive marketing, high costs, and inadequate information received or known about the device.

1.7.1. Lack of Information

The act of purchasing a medical device is based on possible needs that the device will fill and as such, the rationale for choosing a medical device will therefore call for information about the needs and extent to which a device will meet that need. A lack of adequate information may result in a failure to adopt a rational or logical assessment of needs which in turn may form one such barrier when it comes to choosing suitable devices that would aid a positive outcome in health care. Much of general health-care spending is tailored to activities that do not actually focus on improving healthcare or better understanding the advantages among various choices which has led to an overall gap in knowledge. This ultimately leads to a lack of knowledge or high-quality data which in turn may result in incorrect choices being made about certain devices carrying a significant degree of risk. A common example is hospitals that choose to offer cutting-edge technology such as proton therapy which is known to be quite expensive. While this treatment is new and advanced, there is no evidence to suggest improved clinical outcomes as when compared to conventional treatment. These scenarios often stem from a lack of objectivity and information availability. There are several sources of information that can assist health care practitioners. Regulatory bodies (such as the FDA in the United States and the Medicines and Healthcare products Regulatory Agency (MHRA) in the United Kingdom) provide information about medical devices however this information is generally limited to that needed for market approval and concerns around safety and efficacy of performance both, pre-, and post-market. It should be remembered that the main function of regulatory authorities is to ensure that a device complies with statutory requirements for safety, performance, and efficacy.

A second factor limiting information availability is that innovative medical devices are usually evaluated for efficacy in clinical trials. These trials however are too small to identify uncommon complications. In addition, randomized controlled trials tend to be impracticable or even unethical in the case of medical devices. One of the results of limited information availability is that estimates around efficacy or rather, the cost-effectiveness of innovative technologies becomes quite vague. Where pre-market clinical trials are concerned, information availability may also become restricted. This occurs as clinicians participating in a pre-market trial tend to be well trained and alert to possible limitations. This is in direct contrast to what the requirement should be-clinicians participating in trials should be unbiased and unaware of the device's shortcomings.

1.7.2. Changing Technologies

Often fascination with science and technology can blind decision-makers and influence their ability to make objective decisions. In more cases than not, it is the existence of a technology that tends to be more appealing than the actual contribution that the technology has to overall patient care. In recent years, many health care practitioners have become reluctant to purchase expensive medical equipment. In fact, national health systems, private insurers, and the like are now demanding that innovation be linked to economic value and focus on improving health outcomes.

1.7.3. Personal Preferences

As time passes, many health care professionals tend to develop personal preferences for specific brands of medical devices. These preferences are generally for brands that they are familiar with or use often and which are often critical to the success of a clinical procedure. It should also be noted that the actual skill of a clinician themselves when using a medical device can also influence the outcome of a procedure which in turn could affect the overall preference towards a device. In addition, the special relationships that clinicians may develop with manufacturers of brands should also be taken into consideration as these may affect the objectivity and professional integrity of the professional concerned. Often these preferred items may include quite expensive devices that, in turn, may account for more than 50% of a hospital's expenditure. Such devices include implantable pacemakers, stents, and so on. Much of these personal preferences can also influence the choice of medical devices chosen in less developed countries. One such example is students who leave poorer countries to study medicine in developed countries like the United States or Europe. During this period, they may learn to work with expensive and complex medical devices while abroad but when returning home, they may not have the same facilities available to them. Therefore, they may put motivations or pressure for the hospital to purchase expensive equipment without taking into consideration the actual public health benefit it could bring.

1.7.4. Costing

The rising costs associated with healthcare form a cause of concern to a growing number of countries. This is especially the case with countries that have aging populations and as a result, increased demands for healthcare. The majority of industrialized countries have cost-containment measures in place for medicines but very few have such measures in place for medical devices. In addition, low-income countries have their own problems. These include extremely high costs associated with expensive equipment which in turn could strain the health system budget. Furthermore, the cost of acquiring a complex medical device is usually only one portion of the financial burden. Accessories to the device, years of warranty, delivery logistics, installation procedures, and maintenance costs, spares, consumables, and reagents all need to be taken into consideration (Box 1). All these additions can in actual fact, amount to more than 80% of the total cost of the device itself. Common examples of hidden costs include service contracts, spare parts, test equipment, downtime, space requirements, licenses, upgrades, utilities, training, supplies, accessories, and installations.

Inconsistency of pricing can lead to further complications during the procurement process while improper cost management systems can pose overall difficulties for some countries. Often, the pricing between public health care facilities and private health care facilities differs drastically which is an additional factor accounting for inconsistencies.

1.8. SOLUTIONS TO OVERCOMING BARRIERS EXPERIENCED WHEN CHOOSING A MEDICAL DEVICE

While there are a variety of solutions to overcoming respective barriers, this section will focus on the main areas that could be applicable to or involved in any possible solution. This includes rationale decision making, public health needs, cost containments, and improvement of existing marketing practices. Often regulatory practices tend to differ from one country to the next. Much of this diversity originates from differences in historical and cultural practices. One of the main objectives of the GHTF is to merge or harmonize differing regulatory practices. This is achieved via various guidelines and recommendations produced by the GHTF which essentially, when focused on collectively; form a framework that any country can use to align their regulatory systems with that of other countries. Currently, there are few standards against which compliance with reliability, safety, efficiency, and quality of medical devices can be measured. International standards have been developed and put forward by several international organizations such as the International Organization for Standardization (ISO), The International Electrotechnical Commission (IEC), and the

International Telecommunication Union; all of which are now widely used within the medical device community.

1.9. COMMON QUESTIONS THAT COULD BE ASKED WHEN CHOOSING A MEDICAL DEVICE

Rational processes should be employed when making decisions about which medical devices to obtain. These processes should consider public health needs, cost-effectiveness, and overall health outcomes. Common questions include:

- What are the clinical needs that the device is aimed at fulfilling?
- Does the need correspond to accepted health-care practices?
- Will the device improve public health?
- Is the cost of the device justified by the need?
- Will paying the cost of the device divert funds needed for other, higher-priority needs?
- Is it a priority in this particular setting?
- Has evidence of adequate safety and effectiveness been obtained?
- Is the device appropriate to the level of health care (primary, secondary, tertiary) at which it will be used?
- Is the device appropriate to the likely availability of resources for upkeep, maintenance, repair, and other ancillary requirements?
- Is training being provided?
- Is the physical infrastructure adequate?

When choosing a medical device, a health care facility or public health authority should make the decision based on answers to various, relevant questions as identified above. These questions are best answered by a team consisting of several specialists within the health care or scientific fields. These could include individuals such as biomedical engineers, medical staff, nursing staff, and even financial staff; all of whom have relevant knowledge and who are mandated to make choices that focus on the health needs of a target population. The size and composition of the team will differ depending on the size and nature of the health-care facility. Furthermore, clinical guidelines could be more useful. A guideline could take various forms, such as a flowchart or decision algorithm which ideally guides the user through a series of steps or choices that ultimately identifies a list of suitable devices needed for the management of a given disease or disability. These protocols should, ideally, urge readers to consider the required technical skills and supportive infrastructure needed and which is available for safe and effective use of the recommended device. These guidelines or protocols could be used at any level of health system; from district hospitals to provincial hospitals to regional hospitals; as well as for any clinical conditions ranging from mild to severe.

1.10. UNDERSTANDING THE ISO STANDARDS

The ISO has put forward a formal definition of a standard that should be adopted where medical devices are concerned. This definition is as follows:

Standards are documented agreements containing technical specifications or other precise criteria to be used consistently as rules, guidelines, or definitions of characteristics, to ensure that materials, products, process, and services are fit for their purpose.

International standards form the building blocks for harmonized regulatory processes to assure the safety, quality, and performance of medical devices. Regulatory Authorities and various Standards Organizations should encourage and support the development of international standards for medical devices such that they demonstrate compliance with "the Essential Principles (EP) of Safety and Performance of Medical Devices." Those Regulatory Authorities developing new medical device regulations should encourage the use of international standards and also provide a mechanism for recognizing international standards such that they allow for demonstration of compliance with the EP.

International standards serve a range of different purposes. They can:

- Provide reference criteria that a product, process or service must meet;
- Provide information that enhances the safety, reliability, and performance of products, processes, and services;
- Assure consumers about reliability or other characteristics of goods or services provided in the marketplace; and
- Give consumers more choice by allowing the product of one company to be substituted for, or combined with the product of another company.

The increasing globalization of medical devices makes the need and subsequent benefits provided from such standards extremely important more so, on an international level where manufacturing, trade, and communications are concerned. As more and more medical devices are used globally, the need for global communication and as a result international standardization has since increased in an attempt to ensure safety, performance, and consistent quality all round.

1.11. TYPES OF STANDARD SPECIFICATIONS

Standards can establish a wide range of specifications for products, processes, and services. The various types of specifications in standards are outlined below:

- **Prescriptive Specifications:** These specifications necessitate product characteristics such as device dimensions, biomaterials, and test and/or calibration procedures. Prescriptive specifications also determine definitions of terms and terminologies.
- **Design Specifications:** These specifications set out the specific design or technical characteristics of a product.
- **Performance Specifications:** These specifications ensure that a product meets a prescribed test. This includes strength requirements, measurement accuracy, and battery capacity, and so on.
- **Management Specifications:** These specifications set out requirements for the processes and procedures which companies put in places such as quality systems for manufacturing or environmental management systems.

In recent times, the development and application of what are known as "generic management system standards" have become more apparent. The term "generic" ideally refers to standards' requirements that can be applied to any organization regardless of the product it makes or the service it delivers while the phrase "management system" refers to what the organization does to manage its processes. Terms such as "outcome-oriented" standards, "objective" standards, "function-focused" standards and "result-oriented" standards are also used. In simpler terms, these kinds of standards refer to those standards which specify the objectives or ends to be achieved while leaving the methods to the implementers.

1.12. VOLUNTARY AND MANDATORY STANDARDS

Most standards are voluntary however; a standard may be mandated or delegated by a company, professional society, industry, government, or trade agreement. If a standard becomes mandated, it may then be called a regulation and may, or may not, have a legal basis. When a standard is mandated by a government or an international trade agreement, it normally becomes legally obligatory. This is often based on regulations or a law established by the government or the contracts between such international bodies.

The use of voluntary standards came about due to the realization that while regulations generally address the essential safety and performance principles; manufacturers and users still need to know detailed specifications pertaining to specific products. Providing these specifications and detailed requirements, however, presents a mammoth task for regulatory authorities. That being said, a large amount of voluntary standards available or currently being developed meant an alternative way to provide such precise specifications. The use of voluntary standards has many advantages. These include:

- These standards are normally developed by experts who have access to a range if resources in the professional and industrial communities;
- Through existing resources, the government can overcome its own limited resources when providing product-specific technical requirements and characteristics;
- Conformity to standards can be assessed by an accredited third party, for example, a notified body in Europe;
- The use of international standards facilitates harmonized regulatory processes thereby improving global access to new technology;
- It is much easier to update standards than to change regulations as technology advances and changes; and
- Manufacturers have the flexibility to choose appropriate standards or other means to demonstrate compliance with regulatory requirements.

1.13. UNDERSTANDING THE NATIONAL AND INTERNATIONAL STANDARD SYSTEMS

While a country can have many voluntary standard bodies, ideally there is one official national organization which will coordinate and accredit standard development and standard development bodies in the country. This official national organization would have the authority to endorse a document as a national standard while also representing the country in various ISOs. These organizations will differ from country to country, for example, in the United States it is the American National Standards Institute (ANSI), and in Canada, it is the Standards Council of Canada (SCC).

The presence of a standards system not only helps medical device administration but is also important for other industrial and economic developments. Having a standardized infrastructure has now become a basic requirement for successful implementation of policies that ideally aim to improve productivity, market competitiveness, and export capability. The three major international standardization organizations are: (1) the International Organization for Standardization (ISO), (2) the International Electrotechnical Commission (IEC), and (3) the International Telecommunication Union (ITU). Each of these international standardization organizations covers different areas, namely ITU covers telecommunications, IEC covers electrical and electronic engineering, and ISO covers anything else. There are other organizations which are known to produce international standardization documents which are subsequently adopted by ISO/IEC/ITU. These standards are then adopted as if they have been developed in accordance with international consensus criteria.

Links to these three major international standardization organizations are given below:

- The International Organization for Standardization (ISO): www.iso.ch.
- The International Electrotechnical Commission (IEC): www.IEC.ch.
- The International Telecommunication Union (ITU): www.itu.int.

1.14. UNDERSTANDING HOW TO IDENTIFY STANDARDS

Standards are generally named by an alphabetical prefix and a number. The letters indicate the international body that has approved them such as ISO, IEC, ANSI, CAN, EN, and DIN to name a few; while the numbers identify the specific standard and the year in which it was finalized. In certain instances, the standard reference code will give an indication of adoption where standards are equivalent. For example:

ANSI/AAMI/ISO 15223:2000 means the international standard ISO 15223 (established in 2000) adopted by the Association for the Advancement of Medical Instrumentations in the United States, which, in turn, is designated by the American National Standards Institute (ANSI) as an American national standard.

And *UNI EN ISO 9001 indicates an Italian national standard (UNI) which is an adoption of a European standard (EN), which is itself an adoption of the International Standard ISO9001.*

1.15. CLASSIFICATION OF MEDICAL DEVICES

Medical devices are generally divided into different classes based on their use and risk when in contact with a patient. This is done in an attempt to ensure the correct regulations and quality systems can be applied with the ultimate goal of ensuring safe and efficient use in or on patients. Some countries have their own separate classification system for different groups of medical devices, such as general devices, active implantable medical devices and *in vitro* diagnostic (IVD) medical devices while other countries utilize one system for classification of all medical devices. Regardless of the country or system used, all classification schemes tend to be risk-based. Having a suitable classification system in place ensures a more appropriate and graduated system of control which allows manufacturers themselves to classify their own devices rather than by one or two general committees, which in effect would take too long. This also enforces that manufacturers should be able to know as early as possible in which class their product falls. In the United States of America, for example, medical devices are classified as class I, II, and III with Class III devices representing the highest risk type device and requiring more overall control. Medical devices regulated with the USA are classified through an FDA classification database and are given a seven-digit number based on the category of the product. On the other hand, in the European Union, general devices are classified as class I, class I sterile, class I measuring, class IIa, class IIb or class III; with class III devices representing the highest risk device. In addition, IVD devices have their own classification system while active implantable devices are not classified at all.

While having such a classification system in place is essential, it should also be noted that due to the vast number of variables involved as well as the changing nature of some of these variables, the mere implementation of a simple set of classification rules solely based on technical features is impossible. As a result, specifically in the EU, the European legislator has since established a classification concept which is based on the potential hazard related to the use and possible failure of various medical devices while at the same time taking into account technology used and any related standards, regulations, and/or health policy considerations. This has resulted in the establishment of a small set of criteria; namely the duration of contact with the body, degree of invasiveness, and local vs. systemic effects which can be combined in various ways to determine the class of the product. In addition, despite these various classification systems or existing rules which aim to classify the vast majority of medical devices, there are however a small number of exceptions or confusing classification situations which may arise. Common examples include borderline devices between two classes, devices that cannot be classified by existing rules due to their unusual nature or even situations in which classification may lead to the wrong conformity assessment level being chosen.

Due to the varying classification systems, the GHTF has developed a recommended classification system in which medical devices are divided into classes namely A, B, C, and D with class D representing the highest risk class. This global scheme is however a recommendation to regulatory authorities (Table 1.8).

Table 1.8. Proposed General Classification System for Medical Devices (GHTF, 2006)

Class	Risk Level	Device Examples
A	Low risk	Surgical retractors/tongue depressors
B	Low-moderate risk	Hypodermic needles/suction equipment
C	Moderate-high risk	Lung ventilator/bone fixation plate
D	High risk	Heart valves/implantable defibrillator

Figure 1.4 shows an overall illustration of increasing levels of regulatory requirements as the device risk class increases. These regulatory controls may include (1) operation of a quality system which is recommended for all medical devices; (2) technical data; (3) product testing using in-house or independent resources; (4) documentation of clinical evidence to support the manufacturer's claims; (5) the need for and frequency of independent external audit of the manufacturer's quality system; and (6) independent external review of the manufacturer's technical data.

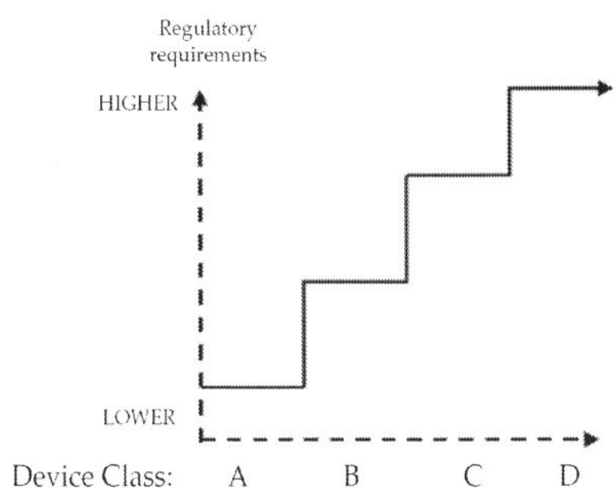

Figure 1.4. Illustration of various regulatory controls as per increasing risk class of medical devices.

1.16. FACTORS INFLUENCING DEVICE CLASSIFICATION

Several factors are known to influence medical device classification. These range from the duration of contact which the device has with the body to the overall degree of invasiveness to whether the medical device is intended to have a biological or systemic effect on the patient. In some instances, the way in which a manufacturer defines the devices' intended use may result in two or more classification rules being applicable to the device itself. In this instance, the highest level of classification will apply. If one medical device is intended to be used together with another medical device that may or may not be from the same manufacturer then the classification rules should apply separately to each of the devices involved; for example, a general-purpose suture and a suture holder. In some instances, medical devices may be grouped and packaged together as per a manufacturer's requirements and in this instance, classification of the medical device assemblage will depend on the manufacturers purpose in packaging and marketing of such medical device combinations as when compared to individually packaged devices. For example, if the combination of devices together results in an overall product that is intended by the manufacturer to meet a specific purpose that is different from that of the individual medical devices that make up the combination, then the combination itself becomes a new medical device in its own right and should be classified according to the new intended use. However, on the other hand, if the medical device combination is for the convenience of the user but does not change the intended uses of the individual medical devices that make up that combination then the classification allocated to that assemblage is at the level of the highest classified device within that group. On the other hand, one or more medical devices within that assemblage of devices have yet to comply with all relevant regulatory requirements, then the combination should be classified based on the intended use of the device.

It is also important to remember that medical accessories that have been intended specifically to be used together with a 'parent' medical device such that it enables that medical device to achieve its intended purpose, should be subjected to all the GHTF guidance documents that apply to medical devices to ensure proper classification and application of relevant regulatory rules. These include, for example, GHTF documents around EP for Safety and Performance, post-market surveillance, medical device classification, and so on. In many instances, these medical device accessories may be classified as though they form actual medical devices in their own right.

In other instances, while most software required for certain medical device operation may be incorporated into the medical device itself, some are not. If this standalone software falls within the scope of what is defined to be a medical device, then this software should be classified as follows:

- if the software drives or influences the use of a separate medical device, it should be classified according to the intended use of the combination.
- if the software is independent of any other medical device, it should be classified in its own right using the rules defined below (taken from Section 8 of GHTF/SG1/N15:2006 Principles of Medical Device Classification, Study Group 1 as endorsed by The Global Harmonized Task Force, 2006); and
- if the standalone software can be defined as being an active device.

1.17. RESPONSIBILITIES OF MANUFACTURERS AND COMMON EXAMPLES OF MEDICAL DEVICES

Medical device manufacturers must ensure that they design devices that are safe, and which perform as intended throughout the entire life cycle of the product. All medical devices have fundamental design and manufacturing requirements which are put in place to ensure they are safe. All medical devices regardless of the class possess a certain degree of risk associated with it and while this is the case, many of the problems cannot be detected until the in-depth market-based experience is gained. More simply put, the safety and risk of the device are judged when that device fails to respond in a manner that it should have, or rather when a device functions in a manner that was not predictable at the time of use or due to unique conditions in the patient itself. For many manufacturers, the current approach is to carry out a suitable risk assessment which involves determination or rather, estimation of the potential of a device becoming a hazard that could subsequently lead to additional problems or harm to end-users.

Clinical effectiveness is a fairly good indicator of how a device performs, thus if a device is said to be clinically effective, it means it produces the effect intended by the manufacturer relative to the medical condition the manufacture claims it will treat. It is much easier and more accurate to measure performance rather than clinical effectiveness. Performance may include technical functions and is very closely linked to the safety of a device, as any medical device that does not perform well could lead to serious clinical safety issues for a patient. For this reason, safety, and performance of medical devices tend to be considered closely together.

Table 1.9. Examples of the Purpose, Place of use and Users of Common Medical Devices

	Medical Devices			
Context of Use	**Preventative**	**Diagnostic**	**Therapeutic**	**Assistive**
Main users	Health-care professionals or healthy individuals	Health-care professionals or patients	Health-care professionals or patients	Individuals or health-care professionals
Examples of medical devices used in healthcare facilities	Surgical gloves, sterilization equipment, disinfectants.	Laboratory diagnostic tests, X-ray equipment, MRI, electrocardiogram, stethoscopes, blood gas analyzers, endoscopes, tongue depressors, reflex hammers.	Orthopedic implants, surgical equipment, pacemakers, stents, infusion pumps, ventilators, sterile dressings, laparoscopes	Traction devices, patient hoists, hospital beds, operating tables, prostheses

Examples of medical devices used in homes	Condoms, gloves, pessary.	Pregnancy tests, blood glucose tests, blood pressure meters, telemedicine, cardiac monitoring.	Infusion pumps, dialysis equipment, oxygen supply systems, syringes.	Crutches, wheelchairs, spectacles, eye lenses, hoists.

CHAPTER 2

What is Biocompatibility?

CONTENTS

2.1. WHAT IS BIOCOMPATIBILITY?

The word 'biocompatibility' refers to the compatibility or interaction between a particular type of material or medical device and the tissues and physiological processes of a living system or patient that may come into with that material or device itself, i.e., it is a measurement of how compatible a device is with a biological system. The main purpose of performing biocompatibility testing is to determine the fitness of a device for human use, and to see if the device may have any potentially harmful physiological effects on the body. Biocompatible materials do not produce a toxic or immunological response when exposed to the body or related bodily fluids.

Biocompatibility of a medical device depends on several factors such as:

- The chemical and physical nature of a device's components;
- The type of patient tissue that will be exposed to the device; and
- The duration of exposure.

In addition, while a particular device is in use, it is possible that substances may leach off into adjacent tissues. These substances are better known as leachables or extractables and in certain instances may not be biologically safe. For this reason, having a final finished medical device as well as the component parts tested for biocompatibility is essential.

Biocompatibility is often investigated using a range of techniques such as analytical chemistry, *in vitro* or *in vivo* tests, and related animal models. Ideally, *in vitro* testing may be conducted first and once completed, *in vivo* biological testing can be done, but the sequence is often based on the device's intended uses. Testing can range from skin irritation testing to hemocompatibility and implantation testing with turnaround tests ranging from three weeks to several months. In fact, subchronic or chronic implantation testing can take even longer.

2.2. DO MEDICAL DEVICE COMPANIES REALLY NEED BIOCOMPATIBILITY TESTING DATA?

Biocompatibility testing data is necessary for medical devices that come into contact with a patient. For this reason, such testing now forms a critical part of the regulatory approval process as even the most advanced designed products can in some cases, produce unintended complications if the device components or even the final finished products are not compatible with living systems. For this reason, manufacturers must consider both corporate regulatory goals and compliance risks when assessing the safety of the devices they place on the market to avoid any potential future product liabilities.

The ISO 10993 Biological Evaluation of Medical Devices series of standards are recognized by all major regulatory bodies as the standard to be used when selecting those biological tests necessary for assessing the safety of a medical device. Ideally, the biological evaluation of any material or medical device intended for use in humans should form part of a structured biological evaluation plan (BEP), together with a properly drawn up risk management process. The risk management process should focus on identifying any biological hazards, estimating associated biological risks, mitigating them, and determining their acceptability (Table 2.1).

Table 2.1. List of Standards in the ISO10993 Series

Standard	Title	Description
ISO 10993-1	Biological evaluation of medical devices Part 1: Evaluation and testing within a risk management process	• The general principle governing the biological evaluation of medical devices within a risk management process; • The general categorization of devices based on the nature and duration of their contact with the body; • The evaluation of existing relevant data from all sources; • The identification of gaps in the available data set on the basis of a risk analysis; • The identification of additional data sets necessary to analyze the biological safety of the medical device; • The assessment of the biological safety of the medical • device.
ISO 10993-2	Biological evaluation of medical devices Part 2: Animal welfare requirements	The standard specifies the minimum requirements to be satisfied to ensure and demonstrate that proper provision has been made for the welfare of animals used in animal tests to assess the biocompatibility of materials used in medical devices.

ISO 10993-3	Biological evaluation of medical devices Part 3: Tests for genotoxicity, carcinogenicity, and reproductive toxicity	The standard specifies strategies for hazard identification and tests on medical devices for the following biological aspects: genotoxicity, carcinogenicity, and reproductive and developmental toxicity.
ISO 10993-4	Biological evaluation of medical devices Part 4: Selection of tests for interactions with blood	The standard provides general requirements for evaluation of the interactions of medical devices with blood. It describes: • A classification of medical and dental devices that are intended for use in contact with blood, based on the intended use and duration of contact as defined in ISO 10993-1; • The fundamental principle governing the evaluation of the interaction of devices with blood; • The rationale for the structured selection of tests according to specific categories, together with the principle and scientific basis of these tests. Detailed requirements for testing cannot be specified because of limitations in the knowledge and precision of tests for interactions of devices with blood. ISO 10993-4 describes biological evaluation in general terms and may not necessarily provide sufficient guidance for test methods for a specific device.
ISO 10993-5	Biological evaluation of medical devices Part 5: Tests for *in vitro* cytotoxicity	The standard describes test methods to assess the *in vitro* Cytotoxicity (cytocompatibility) of medical devices. These methods specify the incubation of culture cells in contact with a device and/or extracts of a device either directly or through diffusion. These methods are designed to determine the biological response of mammalian cells *in vitro* using appropriate parameters.

ISO 10993-6	Biological evaluation of medical devices Part 6: Tests for local effects after implantation	The standard specifies test methods for the assessment of the local effects after implantation of biomaterials intended for use in medical devices. ISO 10993-6:2007 applies to materials that are: • Solid and non-biodegradable; • Degradable and/or resorbable; • Non-solid, such as porous materials, liquids, pastes, and particulates. ISO 10993-6 may also be applied to medical devices that are intended to be used topically in clinical indications where the surface or lining may have been breached, in order to evaluate local tissue responses.
ISO 10993-7	Biological evaluation of medical devices Part 7: Ethylene oxide sterilization residuals	The standard specifies allowable limits for residual ethylene oxide (EO) and ethylene chlorohydrin (ECH) in individual EO-sterilized medical devices, procedures for the measurement of EO and ECH, and methods for determining compliance so that devices may be released. EO-sterilized devices that have no patient contact (e.g., *in vitro* diagnostic devices) are not covered by ISO 10993-7:2008.
ISO 10993-8:2001	Biological evaluation of medical devices Part 8: Selection of reference materials (withdrawn)	The standard gives guidance on the selection and qualification of reference materials for biological tests.
ISO 10993-9:2010	Biological evaluation of medical devices Part 9: Framework for identification and quantification of potential degradation products	The standard provides general principles for the systematic evaluation of the potential and observed biodegradation of medical devices and for the design and performance of biodegradation studies. ISO 10993-9 considers both non-resorbable and resorbable materials.

ISO 10993-10	Biological evaluation of medical devices Part 10: Tests for irritation and skin sensitization	The standard describes the procedure for the assessment of medical devices and their constituent materials with regard to their potential to produce irritation and skin sensitization. ISO 10993-10:2010 includes: • Pre-test considerations for irritation *in silico* and *in vitro* methods for dermal exposure; • Details of *in vivo* (irritation and sensitization) test procedures; • Key factors for the interpretation of the results. Instructions are given for the preparation of materials specifically, in relation to the above tests and several special irritation tests are described for application of medical devices in areas other than skin.
ISO 10993-11	Biological evaluation of medical devices Part 11: Tests for systemic toxicity	The standard specifies requirements and gives guidance on procedures to be followed in the evaluation of the potential for medical device materials to cause adverse systemic reactions.
ISO 10993-12	Biological evaluation of medical devices Part 12: Sample preparation and reference materials.	The standard specifies requirements and procedures to be followed in the preparation of samples and the selection of reference materials for medical device testing in biological systems in accordance with one or more parts of ISO 10993. Specifically, ISO 10993-12 addresses the following: • Test sample selection; • Selection of representative portions from a device; • Test sample preparation; • Experimental controls; • Selection of, and requirements for, reference materials; • Preparation of extracts. ISO 10993-12 is not applicable to live cells, but can be relevant to the material or device components of combination products containing live cells.

ISO 10993-13	Biological evaluation of medical devices Part 13: Identification and quantification of degradation products from polymeric medical devices	ISO 1099313:2010 provides general requirements for the design of tests in a simulated environment for identifying and quantifying degradation products from finished polymeric medical devices ready for clinical use.
ISO 10993-14	Biological evaluation of medical devices Part 14: Identification and quantification of degradation products from ceramics	Identification and quantification of degradation products from ceramics.
ISO 10993-15	Biological evaluation of medical devices Part 15: Identification and quantification of degradation products from metals and alloys	Identification and quantification of degradation products from metals and alloys.
ISO 10993-16	Biological evaluation of medical devices Part 16: Toxicokinetic study design for degradation products and leachables	Toxicokinetic study design for degradation products and leachables.
ISO 10993-17	Biological evaluation of medical devices Part 17: Establishment of allowable limits for leachable substances	Establishment of allowable limits for leachable substances.
ISO 10993-18	Biological evaluation of medical devices Part 18: Chemical characterization of materials	Chemical characterization of materials.
ISO/TS 10993-19	Biological evaluation of medical devices Part 19: Physico-chemical, morphological, and topographical characterization of materials	Physico-chemical, morphological, and topographical characterization of materials.
ISO/TS 10993-20	Biological evaluation of medical devices Part 20: Principles and methods for immunotoxicology testing of medical devices	Principles and methods for immunotoxicology testing of medical devices.

ISO/TR 10993-22	Biological evaluation of medical devices Part 22: Guidance on nanomaterials	Guidance on nanomaterials.

Before a device can enter the market, manufacturers need to determine which biocompatibility tests should be performed in order to determine the potential toxicity resulting from a particular device with the body. This is based on the overall intended use of the device in question. ISO 10993-1 has listed the tests to be performed based on the area of contact between the medical device itself and the patient, as well as its duration of contact (Table). These tests should ideally show that the device materials are not, either directly or through the release of any leachables from material constituents, producing any local or systemic adverse effects, are not carcinogenic or producing any adverse reproductive and/or development effects.

The biological evaluation of any material or medical device intended for use in humans form part of what is known as a structured biological evaluation program which is prepared within a risk management process in accordance with ISO 14971 which is the standard for application of risk management to medical devices. A structured biological evaluation program is set out in accordance with ISO 10993 (Figure 2.1).

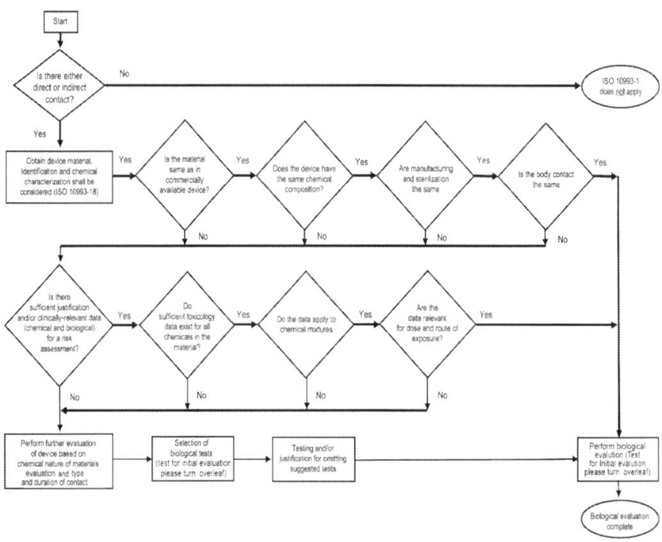

Figure 2.1. Flow chart showing a systematic approach to the biological evaluation of medical devices as part of a risk management process.

As per ISO 10993, the choice of tests and the data required in a biological evaluation shall take into consideration the chemical composition of the materials, the conditions of exposure as well as the nature, degree, frequency, and duration of exposure of the medical device or its constituents to the body, enabling the device to be categorized which then facilitates the selection of appropriate tests (Table 2.2).

Table 2.2. Biological Evaluation of Medical Devices for Initial Evaluation

Device categories			Initial evaluation								Biological Effect	
	Body contact	Contact duration	Cytotoxicity	Sensitization	Irritation or intracutaneous reactivity	Systemic toxicity (acute)/pyrogenicity	Subchronic toxicity	Genotoxicity	Implantation	Hemocompatibility	Chronic toxicity	Carcinogenicity
Surface devices	Skin	A	•	•	•							
		B	•	•	•							
		C	•	•	•							
	Mucosal Membranes	A	•	•	•							
		B	•	•	•	o	o		o			
		C	•	•	•	o	•	•	o		o	
	Breached/ compromised surface	A	•	•	•	o						
		B	•	•	•	o	o		o			
		C	•	•	•	o	•	•	o		o	
External communication devices	Blood path indirect	A	•	•	•	•				•		
		B	•	•	•	•	o			•		
		C	•	•	o	•	•	•	o	•	o	o
	Tissue/bone/dentin communicating	A	•	•	•	o						
		B	•	•	•	•	•	•	•			
		C	•	•	•	•	•	•	•			
	Circulating blood	A	•	•	•	•		o		•		
		B	•	•	•	•	•	•	•	•		
		C	•	•	•	•	•	•	•	•	o	o
Implant devices	Bone/tissue	A	•	•	•	o						
		B	•	•	•	•	•	•	•			
		C	•	•	•	•	•	•	•		o	o
	Blood	A	•	•	•	•	•		•	•		
		B	•	•	•	•	•	•	•	•		
		C	•	•	•	•	•	•	•	•	o	o

A. limited exposure (5-24 hours); B. prolonged exposure (24 hours-30 days); C. permanent contact (> 30 days)

2.3. CATEGORIZATION OF MEDICAL DEVICES

As per ISO 10993-1, medical devices shall be categorized according to the nature of body contact (surface device external communicating device and/ or implant device) and the duration of body contact (limited: < 24 hours; prolonged: 24 h to 30 days and permanent: > 30 days) as per the intended use of the device. By categorizing the device in this way, it facilitates the selection of appropriate testing.

2.3.1. Categorization by Nature of Body Contact

Medical devices can be categorized by the nature of body contact into four main categories, namely-non-contacting, surface-contacting, externally communicating, and implant medical devices. Non-contacting medical devices include those devices or device components that have no direct or indirect contact with the body. Surface contacting devices include devices that come into contact with the skin, mucosal membranes, and breached or compromised surfaces while externally communicating devices those that are in contact with the blood path, tissue, bone, dentin, and circulating blood. Implant devices are those manufactured to replace a missing biological structure, support a damaged biological structure or to enhance an existing biological structure. Non-contact devices are those devices that do not contact the patient's body directly or indirectly. Common examples include *in vitro* diagnostic (IVD) devices. Regulatory agencies rarely require biocompatibility testing for such devices (Table 2.3).

Table 2.3. Overview of Different Device Categories with Definitions and Examples

Device Categories		**Definition**	**Examples**
Surface device	Skin	Devices that contact intact skin surfaces only.	Examples include: Electrodes External prostheses Fixation tapes Compression bandages Monitors of various types
	Mucous membrane	Devices communicating with intact mucosal membranes.	Examples include: Contact lenses Urinary catheters Intravaginal devices Intra-intestinal devices Stomach tubes Sigmoidoscopes Colonoscopes Gastroscopes Endotracheal tubes Bronchoscopes Dental prostheses Orthodontic devices IUD's

	Breached or compromised surfaces	Devices that contact breached or otherwise compromised external body surfaces.	Examples include: Ulcer dressings Burn dressings Granulation tissue dressings Healing devices Occlusive patches
External communicating devices	Blood path indirect	Devices that contact the blood path at one point and serve as a conduit for entry into the vascular system.	Examples include: solution administration sets Extension sets Transfer sets Blood administration sets
	Tissue/bone/dentin communicating	Devices communicating with tissue, bone, and pulp/dentin system.	Examples include: Laparoscopes Arthroscopes Draining systems Dental cements Dental filling materials Skin staples
		This category also includes devices which contact internal tissues (rather than blood contact devices).	Examples include: Surgical instruments and related accessories
	Circulating blood	Devices that contact circulating blood.	Examples include: Intravascular catheters Temporary pacemaker electrodes Oxygenators Extracorporeal oxygenator tubing and accessories Hemo-adsorbents Immunoabsorbents
Implant device	Tissue/bone	Devices principally contacting bone.	Examples include: Orthopedic pins Plates Replacement joints Bone prostheses Cements Intraosseous devices

		Devices principally contacting tissue and tissues fluid.	Examples Include: Pacemakers Drug supply devices Neuromuscular sensors and stimulators Replacement tendons Breast implants Artificial larynxes Subperiosteal implants Ligation clips
	blood	Devices principally contacting blood.	Examples include: pacemaker electrodes Artificial arteriovenous fistulae Heart valves Vascular grafts Stents Internal drug delivery catheters Ventricular assist devices

2.3.2. Biological Effect

Determining the biological effect is probably one of the most critical steps in a biocompatibility evaluation. The ISO 10993 biocompatibility matrix not only categorizes devices based on the type and duration of body contact but also presents a list of potential biological effects. For each biological effect category identified in the ISO 10993 test matrix, a rationale for the testing strategy should be documented.

2.3.3. Cytotoxicity

Cytotoxicity testing or cell culture assays are used to assess the biocompatibility (or toxicity or irritancy potential) of a material or test extract through the use of isolated cells *in vitro*. There are three cytotoxicity tests that are commonly used or medical devices, namely:

- The direct contact test;
- The agar diffusion assay; and
- The MEM elution assay.

The direct contact test is usually recommended for low-density materials. This test involves a piece of test material that is placed directly into cells growing on a culture medium. The cells are then incubated. During incubation, leachable chemicals from the test material can diffuse into the culture medium and contact the cell layer. Cytotoxicity or reactivity of the test sample is indicated by malformation, degeneration, and lysis of cells around the test material.

The agar diffusion assay is generally used for high-density materials and involves a thin layer of nutrient supplemented agar being placed over cultured cells. The test material, or an extract of the test material is placed on top of an agar layer and the cells are incubated. If the sample is cytotoxic, a zone of malformed, degenerative, or lysed cells under and/or around the test material will form.

The MEM Elution assay uses different extracting media and extraction conditions to test devices based on their actual use conditions, or in some instances to exaggerate those conditions. Extracts can be titrated to yield a semi-quantitative measurement of cytotoxicity. After preparation, the extracts are transferred onto a layer of cells and incubated. Following incubation, the cells are examined under a microscope for malformation, degeneration, and lysis.

2.3.4. Sensitization Assays

These assays help to determine whether a material consists of chemicals that cause adverse local or systemic effects after repeated or prolonged exposure. These allergic or hypersensitive reactions involve immunological mechanisms. Sensitization assays may be performed using specific chemicals from the test material, the actual test material itself or extracts of the test material.

The Guinea Pig Maximization Test (GPMT) (also known as Magnusson-Kligman method) is usually recommended for externally communicating devices or those devices that have internal contact with the body or body fluids. In this study, the test material is mixed with complete Freund's adjuvant (CFA) to enhance the skin sensitization response. *Further explanation of the Guinea Pig Maximization Test is provided in Section 3.2.2.*

The closed patch test involves multiple topical doses and is recommended for devices that will contact only unbroken skin. *Further explanation of the Closed Patch Test is provided in Section 3.2.2.2.*

2.3.5. Irritation Tests

Irritation tests determine an estimate of the local irritation potential of devices, materials, or extracts. This is done using sites such as skin or mucous membranes, usually in an animal model. The route of exposure whether it be skin, eye, or mucosa as well as the duration of contact during the test should be similar to the anticipated clinical use of the device. In many cases, testing will attempt to exceed predicted exposure conditions so as to establish a margin of safety for patients.

The Intracutaneous Test involves extracts of the test material which are injected intradermally. The injection sites are then scored for redness and swelling. This type of testing is usually recommended for externally communicating devices or those devices with internal contact within the body or body fluids. It reliably detects the potential for local irritation due to chemicals that may be extracted from a biomaterial. *Further explanation of the Intracutaneous Test is provided in Section 3.1.1.*

The Primary Skin Irritation test should be considered for topical devices that have external contact with intact or breached skin. The test material or an extract of the test material is applied directly to intact and abraded sites on the skin of a rabbit. After a 24-hour exposure, the material is removed and the sites are scored for erythema and edema. *Further explanation of the primary skin irritation test is provided in Section 3.1.2.*

The Mucous Membrane Irritation Tests are recommended for externally communicating devices that will have contact with intact natural channels or tissues. These studies often use extracts rather than the material from the medical device itself. Some common procedures include vaginal and eye irritation studies.

2.3.6. Acute Systemic Toxicity

This range of tests aims to detect leachables that produce systemic toxic effects. The extracts of the test material and negative control blanks are injected into mice (intravenously or intraperitoneally, depending on the extracting media). The mice are observed for toxic signs just after injection and at four other time points. This test is recommended for all blood contact devices but may also be suitable for any other device that comes into contact with internal tissues.

The Material Mediated Pyrogen test evaluates the potential of a material to cause a pyrogenic response, or fever, when introduced into the blood. Lot release testing for pyrogenicity is done *in vitro* using the bacterial endotoxin (LAL) test and must be validated for each device or material. However, for assessing biocompatibility, the rabbit pyrogen test is preferred. This test aims to detect bacterial endotoxins and is sensitive to material-mediated pyrogens that may be found in test materials or extracts.

2.3.7. Sub Chronic Toxicity

These tests are used to determine the potentially harmful effects from long-term or multiple exposures to test materials and/or extracts of test materials during a period of up to 10% of the total lifespan of the test animal. Subchronic tests are required for all permanent devices and should also be considered for those with prolonged contact with internal tissues. Actual use conditions of a medical device need to be taken into account when selecting an animal model for subchronic toxicity and appropriate animal models are determined on a case-by-case basis.

2.3.8. Genotoxicity

Genotoxicity testing involves a set of *in vitro* and *in vivo* tests to detect mutagenic substances. These are substances that can directly or indirectly induce genetic damage by various mechanisms. These genotoxic effects include point mutations along a strand of DNA, damage to the overall structure of the DNA or damage to the structure of the chromosome which contains the DNA. This damage can occur in either somatic or germline cells and increase the risk of cancer or inheritable defects.

These effects can be determined by a variety of tests, each of which aims to detect if DNA damage has occurred at any of the above-mentioned levels. These assays complement one another and are performed as a range of tests. The most common test for mutagenicity is known as the Ames test and detects point mutations by employing use of Salmonella typhimurium. These bacterial strains have been selected for their sensitivity to mutagens.

The Mouse Lymphoma and the HGPRT assays are common procedures for detecting point mutations using mammalian cells. The Mouse Lymphoma assay is also able to detect clastogenic lesions in genes (chromosome damage). Assays for DNA damage and repair include both *in vitro* and *in vivo* Unscheduled DNA Synthesis (UDS).

Cytogenetic assays allow for the direct observation of chromosome damage and include both *in vitro* and *in vivo* methods. These include the

Chromosomal Aberration and the Mouse Micronucleus assays.

The biocompatibility standard guides that an assessment of genotoxic potential for permanent devices and for those with prolonged contact (>24 hours) with internal tissues and blood should be conducted. Extracorporeal devices with limited contact (<24 hours) may require a genotoxicity evaluation. Generally, devices with long-term exposure require an Ames test and two *in vivo* methods, usually the Chromosomal Aberration and Mouse Micronucleus tests. Devices with less critical body contact may be able to be tested using only the Ames test. *Further explanation is provided in Section 3.5.*

2.3.9. Implantation Tests

Implantation studies are used to evaluate the biocompatibility of absorbable and/or non-absorbable materials from medical devices that are in direct contact with living tissue other than skin. Common examples include sutures, surgical ligating clips, and implantable devices. Ideally, these tests are used to evaluate devices which are intended to be implanted for either short-term or long-term periods. Biochemical exchange and cellular and immunologic responses may be assessed in implantation studies through the use of histopathology. *Further explanation is provided in Section 3.6.*

2.3.10. Hemocompatibility

Hemocompatibility studies are used to determine if materials used in blood-contacting devices are compatible with the blood. In theory, all materials are to some degree incompatible with blood because they can either disrupt the blood cells or activate the coagulation pathways. The hemolysis assay is recommended for all devices or device materials except those which contact only intact skin or mucous membranes. This test measures the damage to red blood cells (RBC) when they are exposed to materials or their extracts, and compares it to positive and negative controls (Table 2.4).

rther explanation is provided in section 3.7

Table 2.4. Devices or Components Which Contact Circulating Blood and the Categories of Appropriate Testing (Externally Communicating Devices)

	Test Category				
Device Examples	**Thrombosis**	**Coagulation**	**Platelets**	**Hematology**	**Complement System**

Catheters in place for less than 24 hours	x[a]			X[b]	
Blood monitors	x[a]			X[b]	X[c]
Catheters in place for more than 24 hours	x[a]			x[b]	x
Cell savers		x	x	x[b]	
Devices for absorption of specific substances from blood		x	x	x	x
Donor and therapeutic apheresis equipment and cell separation systems		x	x	x	x
Extracorporeal membrane oxygenator systems, hemodialysis/hemofiltration equipment, percutaneous circulatory support devices	x[a]			x	x
Leukocyte removal filter		x	x	x[b]	x

[a] Thrombosis is an *in-vivo* or *ex-vivo* phenomenon. Coagulation and platelet response are involved in this process. The manufacturer must decide if testing in coagulation and platelet testing are appropriate;

[b] Hemolysis testing only; and

[c] Only for aphaeresis equipment.

Table 2.5. Devices or Components Which Contact Circulating Blood and the Categories or Appropriate Testing-Implant Devices

	Test Category				
Device Examples	**Thrombosis**	**Coagulation**	**Platelets**	**Hematology**	**Complement System**
Annuloplasty rings, mechanical heat valves	x[a]			x[b]	
Intra-aortic balloon pumps	x[a]			x[b]	x
Total artificial hearts, ventricular-assist devices	x[a]			x[b]	
Embolization devices	x[a]			x[b]	x
Endovascular devices	x[a]			x[b]	
Endovascular grafts	x[a]			x[b]	

Implantable defibrillators and cardioverters	X[a]			X[b]	
Pacemaker leads	X[a]			X[b]	X
Prosthetic vascular grafts and patches	X[a]			X[b]	
Stents	X[a]			X[b]	
Tissue heart valves	X[a]			X[b]	
Tissue vascular grafts and patches	X[a]			X[b]	
Vena cava filters	X[a]			X[b]	

[a] Thrombosis is an in-vivo phenomenon. Coagulation and platelet response are involved in this process. The manufacturer must decide if testing in coagulation and platelet testing are appropriate.

[b] Hemolysis testing only.

2.3.11. Carcinogenesis Bioassays

These assays are used to determine the tumorigenic potential of test materials and/or extracts from either a single or multiple exposures, over a period consisting of the total life span of the test system (e.g., two years for rat, 18 months for mouse, or seven years for dog). This type of testing is expensive and can be quite problematic and/or controversial. Often manufacturers negotiate an alternative to full-scale carcinogenicity testing.

2.3.12. Reproductive and Developmental Toxicity Testing

These studies evaluate the potential effects of test materials and/or extracts on fertility, reproductive function, and prenatal and early postnatal development. They are often required for devices with permanent contact with internal tissues.

2.3.13. Pharmacokinetics

Pharmacokinetic studies, also known as Absorption/ Distribution/ Metabolism/Excretion (ADME) studies are used to investigate the metabolic processes of absorption, distribution, biotransformation, and elimination of toxic leachables and potential degradation products from test materials and/or extracts. They are commonly used for bioabsorbable materials or for drug/device combinations.

2.4. HOW TO DETERMINE IF A MANUFACTURER MUST TEST THEIR DEVICE AND WHICH TESTS APPLY

A manufacturer must first determine whether their device requires biocompatibility testing before determining what types of tests they are required to perform. For most devices that have significant tissue contact, some form of biocompatibility testing is required. In most instances, companies arrange for their own biocompatibility studies to be carried out an accredited laboratory. Depending on the type and amount of biocompatibility data available for a particular device, a manufacturer may be able to reduce, the amount testing (and subsequent costs) needed. This includes:

2.4.1. Availability of Biocompatibility Testing Data from Previous Tests or Submissions to Regulatory Bodies

If previous biocompatibility testing data is available, manufacturers can use this or alternatively, they may need to perform some kind of confirmatory testing if significant changes to the device have occurred. These changes include:

- Changes in materials used for device manufacture;
- Changes in the manufacturing processes;
- Changes in the chemical composition of materials used;
- Changes in device intended use or nature of patient contact; and
- Changes in sterilization methods.

2.4.2. Availability of Material/Component Biocompatibility Data from Suppliers

If manufacturers are obtaining certain materials or medical device components from different suppliers, it is important for a manufacturer to check if that supplier has conducted biocompatibility testing on the device. Manufacturers should obtain and screen copies of the original biocompatibility study reports before agreeing to work with that supplier. In addition, manufacturers should attempt to go one step further and ensure that the laboratory conducting the testing is accredited has experienced staff and a strong track record of cGMP/GLP compliance. In addition, manufacturers should perform some confirmatory testing of their own (in addition to the testing done by their suppliers) so as to support and confirm such vendor data.

2.4.3. Analytical Data

Some manufacturers may use analytical data to help demonstrate that a device has a low overall risk of producing a particular biological effect. This takes section 18 of ISO 10993 Standard (Chemical Characterization of Materials) into consideration.

2.4.4. Clinical Data

The availability of clinical data can be used to show and satisfy some biological effects categories from the ISO 10993-1 test selection matrix. This data can be generated from clinical trials of the device in question or from clinical experience with predicate devices or devices containing similar components or materials.

The ISO 10993 series of standards is ideally used for confirmation of the suitability of device materials for its intended application. The first step in this process is chemical characterization of device components with biological testing forming the most critical step in any biocompatibility evaluation. The ISO 10993-1 materials biocompatibility matrix provides a framework that manufacturers can use to design a suitable biocompatibility testing program. In addition, manufacturers, and/or product development and regulatory personnel should consult with an experienced device toxicologist as well as clinical investigators to determine how best to meet the requirements of the materials biocompatibility matrix.

For each biological effect category, the rationale for the testing strategy should be documented. This is especially true when a manufacturer decides not to perform testing for an effect specified by the matrix for their category of devices.

2.4.5. Chemical Characterization

The degree of chemical characterization required for a particular medical device ideally should be determined based on the nature and duration of clinical exposure as well as the data necessary to evaluate the biological safety of the device. In addition, it will also depend on the nature of materials used to design and develop the device. These materials can include liquids, gels, polymers, metals, or biologically sourced material.

Manufacturers should develop a suitable program for chemical characterization. An example is given below:

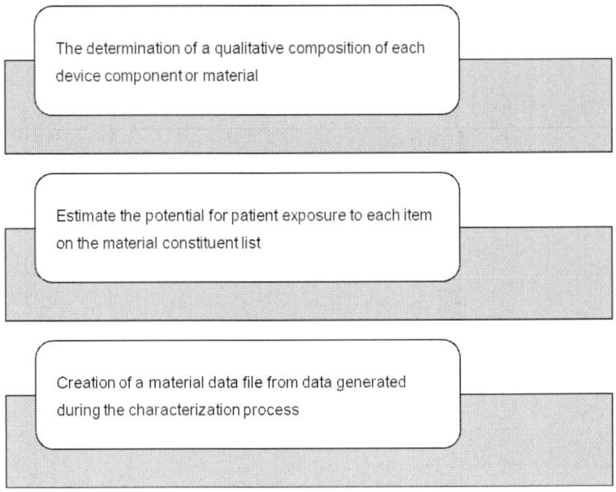

- **Determination of the Qualitative Composition of Each Device Component or Material:** This information can be made available from the material supplier or determined through lab testing. This information can include: matrix identification, information on plasticizers, colorants, anti-oxidants, and other materials that are deliberately added during fabrication of the material; any impurities present and manufacturing materials used such as solvent residues and lubricants.
- **Estimation of the Potential for Patient Exposure:** Manufacturers should use basic literature searches and analyses of toxicological databases to assess the potential for tissue reactivity. Where potentially toxic constituents are considered, lab studies that determine the extractable levels of such constituents should be designed.
- **Creation of a Material Data File Based on Generated Data:** The information generated can be used to create a material data file which in turn can be used as a consistency reference for continuous testing of materials in different lots. This may in turn reduce the need for routine biological testing.

Box 3: Additional Uses on Analytical Characterization Data

Additional uses of analytical characterization data might include:

1. Use in an assessment of the overall biological safety of a medical device;
2. Measurement of the level of any leachable substance in a medical device in order to allow the assessment of compliance with the allowable limit derived for that substance from health-based risk assessment;
3. Judging equivalence of a proposed material to a clinically established material;
4. Judging equivalence of a final device to a prototype device to check the relevance of data on the latter to be used to support the assessment of the former; and
5. Screening of potential new materials for suitability in a medical device for a proposed clinical application.

2.5. KEY CONCEPTS OF MATERIAL CHARACTERIZATION AND ANALYTICAL TESTING OF BIOMATERIALS

Analytical Testing allows for in depth biocompatibility testing of materials that make up medical devices. Learning more about the biocompatibility of the materials making up respective devices as well as their tendency to release leachable matter will help manufacturers to properly assess the risks on *in vivo* reactivity and prevent subsequent toxicology problems arising with the finished devices. In addition, ISO 10993-17 and 10993-18 indicate the requirements for the analytical characterization of device materials and potential leachables. The degree of chemical characterization required should ideally reflect the nature and duration of the clinical exposure and should be determined based on the data needed to evaluate the biological safety of the device. In addition, it will also depend on the nature of materials used.

An example of a strategy for chemical characterization of a device is presented below:

1. **Determine the qualitative composition of each device component or material:** This information should be available from the material supplier. Alternatively, it can be determined through laboratory testing. The list of constituents should include:
    i. The identity of the matrix (i.e., the major component such as the specific polymer, alloy, or metal;
    ii. All plasticizers, colorants, antioxidants, fillers, etc., deliber-

ately added during fabrication of the material;

iii. Impurities such as unreacted monomers and oligomers; and

iv. Manufacturing materials such as solvent residues slip agents, and lubricants.

2. **Estimate the potential for patient exposure for each item on the material constituent list:** Use literature searches of toxicological databases to assess the likelihood of tissue reactivity. For potentially toxic constituents, design, and conduct laboratory studies to determine the extractable levels of those constituents. Use exaggerated conditions of time and temperature, and consider appropriate detection limits. Additional studies may be needed to assess levels of extractables released in actual use conditions.
3. **Data generated from this characterization process can be used to create a material data file:** The information can then be used as a reference for continued testing of device materials to ensure consistency of future production lots. This may in turn reduce the need for routine biological testing.
4. **Additional uses of analytical characterization data might include:**
   i. The data could be used in an assessment of the overall biological safety of a medical device.
   ii. Measurement of the level of any leachable substance in a medical device. This will allow for assessment of compliance with the allowable limit derived for that substance from health-based risk assessment.
   iii. Judging equivalence of a proposed material to a clinically established material.
   iv. Judging equivalence of a final device to a prototype device. This will allow a manufacturer to check the relevance of data on the latter which could be used to support the assessment of the former.
   v. Screening of potential new materials for suitability in a medical device.

Table 2.6. Overview of Some Tests Used in Material Characterization

Group	**Material Characterization Technique**
Traditional extractable material characterization	USP physicochemical tests-plastics
	USP physicochemical test panel for elastomeric closures for injections
	USP polyethylene containers tests-heavy metals and non-volatile residues
	Indirect food additives and polymers extractables (21 CFR Part 177)
	Sterilant residue-ethylene oxide, ethylene chlorohydrin, ethylene glycol
Bulk material characterization	Infrared spectroscopy analysis for identity and estimation of gross composition • Reflectance Spectroscopy • Transmission Spectroscopy
	Atomic absorption spectroscopy (AAS)
	Inductively-coupled plasma spectroscopy (ICP)
	Thermal analysis
Test procedures for extractable materials	UV/visible spectroscopy
	Gas chromatograph
	Liquid chromatography
	Infrared spectroscopy (IR)
	Mass spectrometry
	Residual solvents
	Atomic absorption spectroscopy (AAS
	Inductively-coupled plasma spectroscopy(ICP)
Surface characterization	Scanning electron microscopy (SEM)
	Energy-dispersive x-ray analysis (EDX)
	IR reflectance spectroscopy

2.6. BIOMATERIALS AND MEDICAL DEVICES

Biomaterials can be derived from various sources, most commonly either from nature or they are synthesized in the laboratory. Often a variety of chemical approaches are utilized. These materials include metallic components, polymers, ceramics, or composite materials; all of which are often used and/or adapted for some form of a medical application. In this way, it either augments or replaces a natural function. These functions may be benign such as those used in heart valves or bioactive, such as hip implants. Common examples of biomaterials and related applications are listed in Table 2.7.

Table 2.7. Overview of Various Biomaterials and Related Applications

Material Characterization	**Applications**
Silicone rubber	Catheters, tubing
Dacron	Vascular grafts
cellulose	Dialysis membranes
Poly (methyl methacrylate)	Intraocular lenses, bone cement
polyurethanes	Catheters, pacemaker leads
hydrogels	Ophthalmological devices, drug delivery
Stainless steel	Orthopedic devices, stents
Titanium	Orthopedic and dental devices
Alumina	Orthopedic and dental devices
Hydroxyapatite	Orthopedic and dental devices
Collagen (reprocessed)	Ophthalmologic applications, wound dressings

Table 2.8. Advantages and Disadvantages of Various Biomaterials

Biomaterials	**Advantages**	**Disadvantages**	**Types of Biomaterials**
polymeric	Easy to make complicated items Tailorable physical and mechanical properties Surface modification Immobilize cell etc. Biodegradable	Leachable compounds Absorb water and proteins etc. Surface contamination Wear and breakdown Biodegradation Difficult to sterilize	PMMA PVC PLA/PGA PE PP PA PTFE PET PUR Silicones

bioceramic	High compression strength Wear and corrosion resistance Can be highly polished Bioactive/inert	High modulus (mismatched with bone) Low strength in tension Low fracture toughness Difficult to fabricate	Alumina Zirconia (partially stabilized) Silicate glass Calcium phosphate (apatite) Calcium carbonate
metallic	High strength Fatigue resistance Wear resistance Easy fabrication Easy to sterilize Shape memory	High modules Corrosion Metal ion sensitivity and toxicity Metallic looking	Stainless steel (316 L) Co-Cr alloys Ti, Al, V Au-Ag-Cu-Pd alloys Amalgam (AgSnCuZn-Hg) Ni-Ti Titanium

2.7. BIOCOMPATIBILITY AND TOXICOLOGY OF BIOMATERIALS

The ability of a material to co-exist and interact with a natural substance is known as biocompatibility. Biomaterials incorporated into medical devices are often implanted into tissues and organs. For this reason, the principles governing the structure of normal and abnormal cells, tissues, and organs as well as the structure and function of normal and abnormal tissues and related mechanisms of disease processes should be clearly understood. When a foreign body is present in the wound site, the reaction sequence of the body is referred to as a foreign body reaction. Often the reaction will differ in intensity and duration depending upon the anatomical site involved and therefore, understanding how a foreign body alters the normal inflammatory response is important.

Ideally, biomaterials should not be toxic and much of the toxicology tests are designed such that it can detect substances that migrate out of these biomaterials. Let's take polymers for example: many low-molecular-weight "leachables" exhibit some level of physiologic activity and cell toxicity. Therefore, we can conclude that a biomaterial should not give off anything from its mass unless it is specifically designed to do so.

Biomaterial associated infections (BAI) is one of the most common complications associated with implantation of any biomaterial regardless of form or function. Often these infections usually involve bacterial colonization and biofilm formation on the biomaterial itself. This makes it difficult to treat with antimicrobials or even standard host defenses. Furthermore, infection of the surrounding tissues also plays an important role in BAI. Often infection may be influenced by the composition and design of the implanted biomaterial.

2.8. MECHANICAL AND PERFORMANCE REQUIREMENTS

The mechanical and performance requirements of biomaterials should also eb taken into consideration as each site with the respective medical device interacts with, will have special requirements with regards to geometry, size, mechanical properties, and bio responses. Often the mechanical and performance requirements originate from physical and/or electrochemical properties of the materials as well as the duration of contact (Table 2.9).

Table 2.9. Overview of Some Biomaterial Properties Required in Design and Development of Devices

Mechanical properties	Hip prosthesis	must be strong and rigid
	Tendon material	must be strong and flexible
	Heart valve leaflet	must be flexible and tough
	Dialysis membrane	must be strong and flexible, but not elastomeric
	Articular cartilage substitute	must be soft and elastomeric
Duration of contact	A catheter may be required for just 3 days	
	Bone plate may fulfill its function in 6 months or longer	
	Leaflet in a heart valve must flex 60 times per minute without tearing for the lifetime of the patient (realistically, at least for 10 or more years)	
	Hip joint must not fail under heavy loads for more than 10 years	
Biophysical properties and other related aspects of performance	The dialysis membrane should have a specified permeability	
	the articular cup of the hip joint must have high lubricity	
	The intraocular lens has clarity and refraction requirements	

2.9. REGULATING BIOMATERIALS

The safety and efficacy of a medical device is imperative and one of the first priorities when design and developing a medical device. To prevent inadequately tested devices and materials entering the market, most countries have developed medical device regulatory bodies. In addition, the International Standards Organization (ISO) has introduced international standards which can be used globally to guide the safety of devices. Often the costs around developing and producing materials that are biologically, and clinically tested are large. Regulated characteristics of some biomaterials can be categorized as Table 2.10.

Table 2.10. Regulated Characteristics of Biomaterials

Regulated Characteristics	Details of Requirement
Compatibility	Tissue reactions
	Change in properties: • Mechanical • Physical • Chemical • Electronics • Electrophysical
Mechanical properties	Elasticity
	Yield stress
	Ductility
	Toughness
	Time-dependent deformation
	Creep
	Ultimate strength
	Fatigue strength
	Hardness
	Wear resistance
manufacturing	Fabrication methods
	Consistency and conformity to all requirements
	Quality of raw materials
	Superior techniques to obtain excellent surface finish or texture
	Capability of material to get safe and efficient sterilization
	Cost of product

2.10. UNDERSTANDING THE DIFFERENCE BETWEEN *IN VIVO* AND *IN VITRO* TESTING

There are two broad categories of testing: *in vitro* testing and *in vivo* testing. Each testing category has its own advantages and disadvantages and understanding the differences between such testing types is important as it offers insight into which method is suitable for particular devices as well as the validity of conclusions drawn. *In vitro* refers to the technique of performing a given procedure in a controlled environment outside of a living organism. One of the disadvantages associated with *in vitro* testing is that these types of tests fail to replicate the precise cellular conditions of an organism. For this reason, many *in vitro* studies often lead to results that do not correspond directly to the circumstances occurring in and around a living organism. *In vivo* testing, on the other hand, refers to testing using a whole, living organism as opposed to a partial or dead organism. Animal studies and clinical trials are common examples of *in vivo* research. This type of testing is often employed over *in vitro* testing because it is more suited for observing the overall effects of a device on a living subject (Table 2.11).

Table 2.11. Overview of the Advantages and Disadvantages of *In Vitro* and *In Vivo* Testing

	Advantages	**Disadvantages**
In vitro testing	Method of choice for manufacturers with many devices and variants to be tested	Difficult to capture interactions between different cell types
	Simpler/easier method of testing because of the ease of growing cell cultures compared with the use of animals	Difficult to simulate the consequences of long-term exposure
	Economical	More expensive
	Avoids the need for ethical approval	Ethical approval is required before testing can begin
	Avoids or decreases the need for laboratory trained personnel who are experienced in animal handling	Requires laboratory trained personnel who are skilled in animal handling
	Experimentally controllable	
	Repeatable	
	Fast	

In vivo testing	Simulates real body conditions	Expensive
	Clinically relevant	Time consuming
		Ethical concerns
		More stringent regulatory factors

2.11. TESTING WITH GLP

As a general rule, all biocompatibility testing should be performed in compliance with Good Laboratory Practice (GLP). Testing is done in accordance with GLP regulations apply to the biological safety studies that have been conducted in support of regulatory submissions. These GLP guidelines or regulations oversee all phases of testing, including preparation and approval of study protocols, monitoring tests in progress, and issuance of final reports, as well as facility and study management and the role, played by the Quality Assurance personnel. It is important for a manufacturer to understand what the regulatory requirements are for the territories into which they wish to launch their medical devices. Testing with GLP is required for Investigational Device Exemption (IDE) and Premarket Approval (PMA) submissions in the United States and also seems to be a requirement for European submissions as well. In practice, studies are usually not rejected for lack of GLP treatment but it does add to the credibility behind testing and manufacturers of device components and materials with GLP testing can use the data in any type of regulatory submission.

Generally, testing with GLP stresses the importance of the following components: resources, characterization, rules, results, and quality assurance.

- **Resources:** Organization, personnel, facilities, and equipment

The Study Director is generally appointed by the test facility management, and assumes full responsibility for the GLP compliance of all activities within the study. He or she must be aware of all events that may influence the quality and integrity of the study even when certain phases of testing are delegated to other test sites. This responsibility is reflected in a signed and dated GLP Compliance Statement which is included in all study reports.

- **Characterization:** Test items and test systems

It is necessary that the Study Director have detailed knowledge about the properties of the test item. This includes the identity of the test item, its potency, composition, stability, and impurity profile. All of these characteristics should be known for the test item and should be provided to the Study Director. This is usually documented in a Certificate of Analysis which should be included in the final report of study results.

- **Rules:** protocols, standard operating procedures (SOPs)

The principal steps of studies conducted in compliance with GLP are described in the study Protocol which must be approved and signed by the Study Director before the study starts. Alterations to the study design can only be made through a formal amendment to the Protocol. Adherence to a Protocol is strictly followed and ensures that the study can be reconstructed at a later point in time.

- **Results:** raw data, final report and archives

Results obtained from a GLP study are interpreted by the Study Director based on the study design and actual conduct of the study. The GLP principles do not include allowance for the Out of Specification (OOS) processes that are commonly employed in evaluation of study results for cGMP processes. Any confounding or contributing factors that could result in misinterpretation of study results can be noted by the Study Director when analyzing the results.

- **Quality Assurance:** independent monitoring of research processes

The Quality Assurance personnel is responsible for ensuring that GLP compliance has been attained throughout the test facility as well as within each individual study. In cases where different phases of a study are conducted at multiple sites, it is required that the additional sites have a functioning Quality Assurance Unit (QAU) and that these off-site QAU units provide assurance in the form of a written report to the Study Director stating that these phases have been conducted according to the protocol, and that they are in compliance with GLP processes.

CHAPTER 3

Understanding Biocompatibility Tests in Greater Detail

CONTENTS

3.1. CYTOTOXICITY AND CELL VIABILITY

Viability levels and proliferation rates of cells often form good indicators of cell health but various physical and chemical agents have an affect on cell health and cellular metabolism, often causing toxicity on cells. This can occur via various mechanisms such as the destruction of cell membranes, prevention of protein synthesis, and so on. Cytotoxicity and cell viability testing are often used for general toxicity screening of such chemicals and form a cheap, reliable, and reproducible method. These methods are useful for testing large numbers of samples and are often based on various cellular functions such as cell membrane permeability, enzymatic activity, cell adherence, ATP production, co-enzyme production, and nucleotide uptake activities. The advantages of such testing include less time, reduced cost, and easy automation making it simple and efficient. A range of cytotoxicity and cell viability assays are currently being used and these can be classified according to measurement types of endpoints such as color changes, fluorescence, and so on. These categories of testing are outlined in Table 3.1.

Table 3.1. Overview of Cytotoxicity and Cell Viability Tests

Classification	Examples of Tests
Dye exclusion	Trypan blue Eosin Congo red Erythrosine B assays
Colorimetric assays	MTT assay MTS assay XTT assay WST-1 assay WST-8 assay LDH assay SRB assay NRU assay Crystal violet assay
Fluorometric assays	Alamar Blue assay CFDA-AM assay
Luminometric assays	ATP assay Real-time viability assay

3.1.1. Dye Exclusion Assays

These assays form some of the simplest and widely used methods available. In dye exclusion assays, viable cells do not absorb dyes and tend to exclude them while dead cells do not. It usually involves a simple staining procedure however can become time-consuming when there are large numbers of samples to analyze. A variety of dyes are commonly used and these include eosin, Congo red, erythrosine B, and trypan blue.

3.1.1.1. Trypan Blue Dye Exclusion Assay

This method is used to determine the number of viable and/or dead cells in a cell suspension and is based on the principle that live cells possess intact cell membranes that exclude the dye while dead cells do not. The assay involves preparing serial dilutions of the test compound and incubating these dilutions with either adherent or nonadherent cells for various time periods. After this step, cells are washed and suspended. The cell suspension is mixed with the trypan blue dye and then visually examined to determine if the cells have taken up or excluded the dye. Viable cells will have a clear cytoplasm whilst dead cells will have a blue cytoplasm. The number of viable cells versus dead cells per unit volume is then calculated using light microscopy and expressed as a percentage when compared to untreated controlled cells (Table 3.2).

Table 3.2. Advantages and Disadvantages of Trypan Blue Dye Exclusion Assay

Advantages	**Disadvantages**
Simple	Cell counting is done using a hemocytometer and counting errors may occur
Inexpensive	Poor dispersion of cells may further lead to inaccurate counts
Good indicator of membrane integrity	Cell dispersion may result in some cells being lost
Dead cells are colored within seconds of exposure Trypan blue dye	Inaccurate dilutions may lead to discrepancies in results

	Presence of air bubbles will affect results
	Difficult to process large numbers of samples concurrently as the exact timing of progressive cytotoxic effects cannot be measured
	Trypan blue cannot be used to distinguish between cells that are healthy and cells that are alive but losing cellular functionality
	Trypan blue is known for having toxic side effects on mammalian cells

3.1.1.2. Erythrosine B Dye Exclusion Assay

Erythrosine B is also known as Erythrosine or Red No. 3 and is commonly used as a food coloring agent. This assay works similarly to that of the trypan blue exclusion assay with added benefits as the erythrosine B dye is an alternative biosafe dye (Table 3.3).

Table 3.3. Advantages and Disadvantages of Erythrosine B Dye Exclusion Assay

Advantages	**Disadvantages**
Simple	Time-consuming
Inexpensive	Labor intensive
Bio safe	Cell counting is done using a hemocytometer and counting errors may occur
Versatile	Poor dispersion of cells may further lead to inaccurate counts
	Cell dispersion may result in some cells being lost
	Inaccurate dilutions may lead to discrepancies in results
	Presence of air bubbles will affect results
	Difficult to process large numbers of samples concurrently as the exact timing of progressive cytotoxic effects cannot be measured

3.1.2. Colorimetric Assays

The basic principle behind the colorimetric assay is the measurement of a biochemical marker to evaluate the metabolic activity of cells. The reagents used in these types of assays develop a color in response to the overall viability of cells. This is then measured using a spectrophotometer. Colorimetric assays can be used for adherent or suspended cell lines are often quick and easy to perform.

3.1.2.1. MTT Assay

The MTT assay or, (3-(4,5-dimethylthiazol-2-yl)-2-5-diphenyltetrazolium bromide) assay is one of the most commonly used colorimetric assays when it comes to determining cytotoxicity or cell viability. This assay aims at focusing on cell viability through determination of mitochondrial functioning of cells by measuring the activity of mitochondrial enzymes such as succinate dehydrogenase. The MTT dye is reduced to a purple formazan by NADH. The product is then measured using a spectrophotometer (Table 3.4)

Table 3.4. Advantages and Disadvantages of MTT Assay

Advantages	**Disadvantages**
Simple and easy to conduct	The MTT formazan is insoluble in water
Safe	Well-to-well error may occur
High reproducibility	More controls needed due to increase in false positives or false negatives
Versatile-can be used to determine cell viability and cytotoxicity	Possible over estimation of cell viability due to increased false positives

3.1.2.2. MTS Assay

The MTS assay, also known as the (5-(3-carboxymethyoxyphenyl)-2-(4-5-dimethyl-thiazoly)-3-(4-sulfophenyl) tetrazolium inner salt assay is based on the conversion of a tetrazolium salt into a colored formazan. This is done by the mitochondrial activity of living cells. The greater the viable number of cells in the culture, the greater the amount of formazan produced. This is then measured with a spectrophotometer at 492 nm (Figure 3.5)

Table 3.5. Advantages and Disadvantages of MTS Assay

Advantages	Disadvantages
Simple and easy to conduct	The level of absorbance measured is often influenced by various factors such as incubation time, cell type, cell number
Accurate/high precision	Proportion of MTS reagent to cells influences measured absorbance level
Rapid indication of toxicity	
Rapid	
Sensitive	
Economic	
Can be used for onsite toxicological assessments	

3.1.2.3. XTT Assay

This assay is based on the tetrazolium salt XTT (2,3-bis(2-methoxy-4-nitro-5-sulphophenyl)-5-carboxanilide-2H-tetazolium, monosodium salt) and is a simple method for measuring proliferation of cells. The assay is based on the ability of cells to metabolically reduce the tetrazolium salt XTT to an orange-colored formazan compound after 2–24 hours of incubation. The orange-colored formazan compound is water-soluble and is measured with a spectrophotometer or ELISA reader (Table 3.6).

Table 3.6. Advantages and Disadvantages of XTT Assay

Advantages	Disadvantages
The assay is carried out in a multi-well plate which allows for the analysis of large numbers of samples	The assay is based on cellular reductive capacity-factors influencing the reductive capacity of cells such as enzymatic regulation, pH, cellular ion concentration and so on will affect the final absorbance reading and ultimately the end result
Rapid and easy to conduct	
Safe to carry out	
High sensitivity/accurate	

3.1.2.4. WST-1 Assay

The WST-1 assay, or the 2-(4-iodophenyl)-3-(4-nitrophenyl)-5-(2,4-disulfophenyl)-SH tetrazolium monosodium salt) cell proliferation assay is designed to measure the relative proliferation rates of cells in culture. The principle behind this assay is based on the conversion of the tetrazolium salt WST-1 into a highly water-soluble formazan salt by mitochondrial dehydrogenase enzymes. This salt is released into the cell culture medium. The reaction involves a color change which is directly proportional to the amount of mitochondrial dehydrogenase in cell culture, allowing of the measurement of cellular metabolic activity. The assay involves the use of multi-well plates in which the WST-1 reagent is added directly. After an incubation period of 30 minutes–4 hours, the plates are read at 450 nm (Table 3.7).

Table 3.7. Advantages and Disadvantages of WST-1 Assay

Advantages	Disadvantages
Easy to carry out	Standard incubation time of 2 hours
Safe high reproducibility	

3.1.2.5. WST-8 Assay

The WST-8 assay, also known as the 2-(2-methoxy-4-nitrophenyl)-3-(4-nitrophenyl)-5-(2,4-disulfophenyl)-2H tetrazolium, monosodium salt) assay is a colorimetric assay that aids determination of viable cell numbers. It is a highly stable and water-soluble salt and is often more sensitive than WST-1. The formazan salt produced is water-soluble which provides an additional advantage as this means the assay does not require an extra step to dissolve the salt (Table 3.8).

Table 3.8. Advantages and Disadvantages of WST-8 Assay

Advantages	Disadvantages
WST-8 is not cell permeable and results in low cytotoxicity	Changes in intracellular metabolic activity will impact the reduction of assay substrates causing an inaccurate reflection on overall cell viability
Results in the production of a water-soluble formazan salt which reduces the need for an additional step during detection	

3.1.2.6. LDH Assay

The LDH or lactate dehydrogenase cytotoxicity assay is a colorimetric assay that measures the stable cytosolic lactate dehydrogenase (LDH) enzyme present. LDH is an enzyme that is normally found within the cell cytoplasm and thus will be released from damaged cells. When cell viability is reduced, the permeability of the membrane increases and results in more of the LDH enzyme being released into the cell culture medium. The released LDH enzyme is measured with by a coupled enzymatic reaction that results in the conversion of a tetrazolium salt into a formazan salt. This results in a red color. The resulting red formazan absorbs maximally at 492 nm and can be measured quantitatively at 490 nm (Table 3.9).

Table 3.9. Advantages and Disadvantages of LDH Assay

Advantages	**Disadvantages**
Reliable	Serum and some other compounds have high inherent LDH activity which may result in high background readings
fast	Assay is limited to serum-free/low serum conditions which reduces the scope of the assay as it can no longer allow determination of cell death under normal growth conditions (this involves the used of 10% fetal calf serum)
Simple to conduct	
Indicator of irreversible cell death as the measurement of LDH can only occur when the cell membrane is damaged	

3.1.2.7. SRB Assay

The SRB or sulforhodamine B assay is a rapid and sensitive colorimetric method that measures the drug-induced cytotoxicity in cell cultures. SRB is a bright pink aminoxanthene dye with two sulfonic groups. The SRB bind to a protein basic amino acid residue in trichloroacetic acid fixed cells to provide a sensitive index of cellular protein. The SRB assay can also be used to evaluate colony formation and colony extinction and can be used in attached and suspended cell cultures (Table 3.10).

Table 3.10. Advantages and Disadvantages of SRB Assay

Advantages	Disadvantages
Simple	Homogenous cell suspension is required as cellular clumping may affect assay performance
Fast	
Sensitive	
High reproducibility	
Less sensitive to environmental fluctuations	
Independent of cellular metabolism	

3.1.2.8. NRU Assay

The NRU (Neutral Red Update) assay is one of the most commonly used colorimetric assays to determine cell cytotoxicity and/or cellular viability. The assay is based on the ability of viable cells to take up the neutral red dye. The dye penetrates cell membranes by nonionic passive diffusion and concentrates in the lysosomes. The dye is then extracted from viable cells using an acidified ethanol solution. Absorbance of the dye is measured using a spectrophotometer at 540 nm. The assay allows for the distinguishing between viable, damaged, and dead cells because neutral red uptake is dependent on the capacity of cells to maintain pH. For example, at physiological pH, the net charge of the dye is zero. This charge enables the dye to penetrate the cell membranes. Inside the lysosomes, there is a proton gradient to maintain a pH lower than that of the cytoplasm and this the dye becomes charged and is retained inside the lysosome. When the cell dies or if the pH gradient is reduced, the dye cannot be retained. For this reason, lysosomal uptake of neutral red dye is a highly sensitive indicator of cell viability (Table 3.11).

Table 3.11. Advantages and Disadvantages of NRU Assay

Advantages	Disadvantages
Can be used to distinguish between viable, damaged or dead cells	Pollutants may affect the assay
Good indicator of lysosomal damage	
quick	
Simple	

3.1.2.9. CVS Assay

The Crystal Violet assay (CVS Assay) is a simple method that can be used to determine cell survival and growth inhibition. Adherent cells detach from cell culture plates during death. The crystal violet dye binds to proteins and DNA of viable cells and thus serves as an indicator of attached and viable cells. As cells lose their adherence during cell death, the overall amount of crystal violet staining occurring is also reduced.

Table 3.12. Advantages and Disadvantages of CVS Assay

Advantages	**Disadvantages**
Simple	Insensitive to changes in cell metabolic activity
quick	Unable to measure cellular proliferation
Versatile	

3.1.3. Fluorometric Assays

Fluorometric assays are easy to perform and can also be used to determine cell viability and cytotoxicity. These assays are much more sensitive than spectro photometric or colorimetric assays and involve the used of a fluorescence microscope, fluorometer, and fluorescence micro plate reader or flow cytometer. These assays are applicable to adherent or suspended cells and simple and easy to use.

3.1.3.1. Alamar Blue (AB) Assay

The AB assay, or Alamar Blue assay is sometimes referred to as the resazurin reduction assay and is based on the conversion of the blue nonfluorescent dye resazurin to the pink resorufin which is fluorescent. This is usually done by mitochondrial enzymes and diaphorases. Resazurin is a suitable dye to monitor viable cell numbers as it is a phenoxazine-3-one dye and cell permeable redox indicator. It acts as an intermediate electron acceptor in the electron transport chain between the final reduction of oxygen and cytochrome oxidase by substituting for molecular oxygen as an electron acceptor. The color of the compound is blue and it is nonfluorescent however the resorufin compound is red and highly fluorescent. Viable cells are able to convert resazurin to resorufinthere by increasing the overall fluorescence and color of the culture medium. The quantity of resorufin produced is directly related to the number of viable cells and this can be measured using a microplate reader fluorimeter equipped with a 560 nm excitation/590 nm emission filter. Depending on the cellular metabolic activity of the cells, the incubation period is usually about 1–4 hours (Table 3.13.)

Table 3.13. Advantages and Disadvantages of Alamar Blue (AB) Assay

Advantages	Disadvantages
Inexpensive assay	There may be some fluorescent interference from test compounds
More sensitive that tetrazolium assays	Direct toxic effects on the cells are possible
Can be coupled with other methods to gain more information about the cytotoxicity mechanisms	

3.1.3.2. CFDA-AM Assay

The CFDA-AM assay or 5-carboxyfluorescein diacetate, acetoxymethyl ester assay can also be used for cytotoxicity testing and serves as an indicator for plasma membrane integrity. The CFDA-AM dye is a nontoxic esterase substrate that can be converted from a membrane permeable, nonpolar nonfluorescent substance to a polar, fluorescent carboxyfluorescein (CF) dye. This conversion indicates the integrity of the plasma membrane since only an intact membrane can maintain cytoplasmic milieu needed to support esterase activity (Table3.14).

Table 3.14. Advantages and Disadvantages of CFDA-AM Assay

Advantages	Disadvantages
Nontoxic to cells	Fluorescent interference from test compounds is possible
Inexpensive assay	
More sensitive that tetrazolium assays	

3.1.3.3. GF-AFC Assay

The GF-AFC assay or protease viability marker assay measures the protease enzyme activity of viable cells in order to determine cell viability. The GF-AFC or glycylphenylalanyl-aminofluorocoumarin substrate is restricted to viable cells and can penetrate viable cells. In these cells, cytoplasmic aminopeptidase activity removes the *gly* and *phe* amino acids to release amino fluoro coumarin (AFC). This produces a fluorescent signal which is directly proportional to the number of viable cells present. This protease activity is rapidly lost in dying cells and for this reason, the protease activity forms a selective marker of the viable cell population (Table 3.15)

Table 3.15. Advantages and Disadvantages of GF-AFC Assay

Advantages	Disadvantages
Relatively nontoxic to cells in culture	Fluorescent interference from test compounds is possible
Long term exposure to the GF-AFC substrate results in little change in cell viability	
Cell population remains viable after the assay and thus can be combined with other assays	

3.1.4. Luminometric Assays

Luminometric assays can also be used for the determination of cell proliferation and cytotoxicity. It serves as a fast and simple option that can be performed in either 96-well or 384-well microtiter plates. One of the biggest advantages of the luminometric assays is the stable luminescent signal produced after the addition of the reagent.

3.1.4.1. ATP Assay

The ATP or adenosine triphosphate assay has been widely used to detect damaged cells. ATP is one of the most important chemical energy reservoirs in cells and is used for a range of cellular processes such a biological synthesis, signaling, transport, and movement of substances. For this reason, cellular ATP forms one of the most sensitive endpoints when measuring cell viability. When cells are damaged and start to lose membrane integrity, they also lose their ability to synthesize ATP and as such, the ATP level of cells decreases.

3.2. SENSITIZATION ASSAYS

Sensitization or hypersensitivity reactions usually occur as a result of repeated or prolonged contact with a chemical substance that interacts with the body's immune system. Often those devices that cause sensitization reactions do so by means of extractable chemicals. Reactions may occur after an individual encounters a material repeatedly (latex gloves) or after continuous, prolonged contact (for example an implant).

The sensitization test as per ISO 10993-10 is an *in vivo* test that evaluates the ability of leachables to cause Type IV hypersensitivity (also known as delayed hypersensitivity). Types IV hypersensitivity is a cell-based immune reaction that results in edema and erythema, better known as swelling and redness. In principle, the sensitization test exposes an animal to the test article extracts, or test article across multiple induction phases, and a single challenge phase. The challenge sites of each animal will be observed for evidence of skin sensitizations reactions 24±2 and 48±2 hours after removal of the test articles/extracts.

There are a range of sensitization tests available, namely:

- Murine Local Lymph Node Assay (LLNA);
- Magnusson/Kligman Test (Maximization Test);
- Buehler Closed Patch Method; and
- *In Vitro* Sensitization.

The most commonly used sensitization test is the Magnusson and Kligman (M & K) Assay which works by inducing and re-exposing a medical device or device extract to determine if repeated exposures result in redness and swelling. The results are then usually graded qualitatively using a scale whereby a zero indicates no visible reaction and a three indicates intense redness and/ or swelling.

3.2.1. Murine Local Lymph Node Assay (LLNA)

The murine local lymph node assay (LLNA) was developed in 1989 and is one of the preferred methods for assessing skin sensitization potential. The basic principle behind LLNA is that those substances suspected of causing a reaction will induce primary proliferation of lymphocytes in the lymph nodes. The test substance is applied topically to mice once daily for three consecutive days and on day 6, the mice are injected intravenously with ^{3}H-thymidine. After sacrificing the mice, the lymph nodes are evaluated for radioactivity. The degree of lymphocyte proliferation will directly correlate with the degree of sensitization and thus serves as a measure of allergenic potency to a particular compound (Figure 3.1).

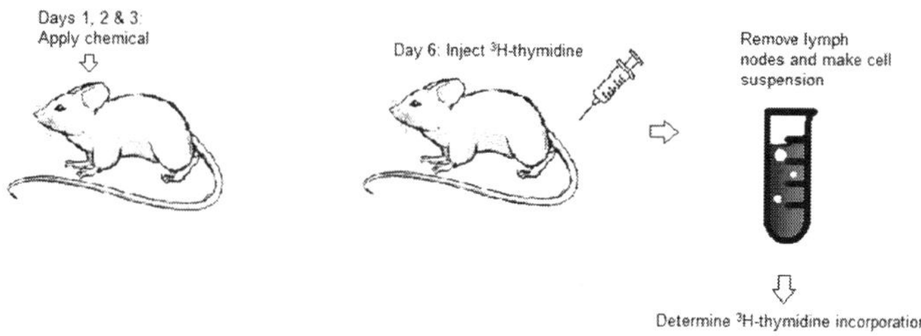

Figure 3.1. Simplistic overview of local lymph node assay.

3.2.2. Guinea Pig Tests

The Guinea Pig tests are commonly used for skin sensitization, in particular two methods namely the guinea pig maximization test (GPMT) and the occluded patch test of Buehler. The purpose of these tests is to identify skin sensitizers effectively.

Several variables should be taken into consideration, and these include:

- Concentrations applied during the induction phase;
- Vehicle(s) used during the induction phase;
- Concentration(s) applied during the challenge phase;
- Vehicle used during the challenge phase;
- Impact of the relative contribution of epicutaneous versus intradermal routes of exposure (GPMT);
- Proportion of guinea pigs with positive challenge reactions;
- Intensity of reactions;
- Kinetics of responses; and
- Dose-response curves.

3.2.2.1. Magnusson/Kligman Test (Guinea Pig Maximization Test)

The Magnusson/Kligman Testor Guinea pig maximization test (GPMT) is an *in vivo* test used to screen for substances that cause human skin sensitization. The test was first proposed by B. Magnusson and Albert Kligman in 1969 and involves exposing test animals intradermally to the test material together with an adjuvant to enhance the immune reaction of the Guinea pig.

Box 4: Description of the Guinea-Pig Maximization Test Method

Number of animals

A minimum of 10 animals is used in the treatment group and at least 5 animals in the control group. When fewer than 20 tests and 10 control guinea pigs have been used, and it is not possible to conclude that the test substance is a sensitizer, then testing additional animals to give a total of at least 20 tests and 10 control animals are strongly recommended.

Dose levels

The concentration of test substance used for each induction exposure should be well-tolerated systemically and should be the highest to cause mild-to-moderate skin irritation. The concentration used for the challenge exposure should be the highest non-irritant dose. The appropriate concentrations can be determined from a pilot study using two or three animals. Consideration should be given to the use of FCA-treated animals for this purpose.

Induction: Intradermal Injections

Day 0-treated group

- Three pairs of intradermal injections of 0.1 ml volume are given in the shoulder region which is cleared of hair so that one of each pair lies on each side of the midline.
    - Injection 1: a 1:1 mixture (v/v) FCA/water or physiological saline;
    - Injection 2: the test substance in an appropriate vehicle at the selected concentration; and
    - Injection 3: the test substance at the selected concentration formulated in a 1:1 mixture (v/v) FCA/water or physiological saline.
- In injection 3, water soluble substances are dissolved in the aqueous phase prior to mixing with FCA. Liposoluble or insoluble substances are suspended in FCA prior to combining with the aqueous phase. The concentration of test substance shall be equal to that used in injection 2.
- Injections 1 and 2 are given close to each other and nearest the head, while injection 3 is given towards the caudal part of the test area.

Day 0-control group

- Three pairs of intradermal injections of 0.1 ml volume are given in the same sites as in the treated animals.
    - Injection 1: a 1:1 mixture (v/v) FCA/water or physiological saline;
    - Injection 2: the undiluted vehicle; and
    - Injection 3: a 50% w/v formulation of the vehicle in a 1:1 mixture (v/v) FCA/water or physiological saline.

Induction: Topical Application

Day 5–7-treated and control groups

Approximately twenty-four hours before the topical induction application, if the substance is not a skin irritant, the test area, after close-clipping and/or shaving is painted with 0.5 ml of 10% sodium lauryl sulfate in Vaseline, in order to create a local irritation.

Day 6–8-treated group

The test area is again cleared of hair. A filter paper (2 x 4 cm) is fully-loaded with test substance in a suitable vehicle and applied to the test area and held in contact by an occlusive dressing for 48 hours. The choice of the vehicle should be justified. Solids are finely pulverized and incorporated in a suitable vehicle. Liquids can be applied undiluted, if appropriate.

Challenge: Topical Application

Day 20–22-treated and control groups

The flanks of treated and control animals are cleared of hair. A patch or chamber loaded with the test substance is applied to one flank of the animals and, when relevant, a patch or chamber loaded with the vehicle only may also be applied to the other flank. The patches are held in contact by an occlusive dressing for 24 hours.

Observations-treated and control groups

- approximately 21 hours after removing the patch the challenge area is cleaned and closely-clipped and/or shaved or depilated if necessary;
- approximately 3 hours later (approximately 48 hours from the start of the challenge application) the skin reaction is observed and recorded according to the grades shown in Table 3.16;
- approximately 24 hours after this observation a second observation (72 hours) is made and once again recorded.

Blind reading of test and control animals is encouraged.

Table 3.16. Magnusson and Kligman Grading Scale for the Evaluation of Challenge Patch Reactions for the Guinea-Pig Maximization Test Method

0	No visible change
1	Discrete or patchy erythema
2	Moderate and confluent erythema
3	Intense erythema and swelling

Rechallenge

If it is necessary to clarify the results obtained in the first challenge, a second challenge (i.e., a rechallenge), where appropriate with a new control group, should be considered approximately one week after the first one. A rechallenge may also be performed on the original control group.

Clinical observations

All skin reactions and any unusual findings, including systemic reactions, resulting from induction and challenge procedures should be observed and recorded. Other procedures, e.g., histopathological examination, the measurement of skin fold thickness, may be carried out to clarify doubtful reactions.

Ref: OECD Test Guideline 406.

3.2.2.2. Buehler Closed Patch Method

The Buehler Closed Patch Test is an *in vivo* skin sensitization test that was first proposed by Edwin Vernon Buehler in 1965. It is a non-adjuvant test that involves exposing Guinea Pigs to a high dose of the test substance and a challenge dose which is the highest dose of the substance that does not cause irritation. The Buehler Closed Patch Method is suitable for screening the sensitization potential of chemicals prior to human exposure. This test method is suitable for strong, moderate, and most weak sensitizers.

Box 5: Description of the Buehler Test Method

Number of animals

- A minimum of 20 animals is used in the treatment group and at least 10 animals in the control group.

Dose levels

- The concentration of test substance used for each induction exposure should be the highest to cause mild irritation. The concentration used for the challenge exposure should be the highest non-irritating dose. The appropriate concentration can be determined from a pilot study using two or three animals.
- For water soluble test materials, it is appropriate to use water or a dilute non-irritating solution of surfactant as the vehicle. For other test materials 80% ethanol/water is preferred for induction and acetone for challenge.

Induction: Topical application

Day 0 – treated group

- One flank is cleared of hair (closely-clipped). The test patch system should be fully loaded with test substance in a suitable vehicle (the choice of the vehicle should be justified; liquid test substances can be applied undiluted, if appropriate). The test patch system is applied to the test area and held in contact with the skin by an occlusive patch or chamber and a suitable dressing for 6 hours.

- The test patch system must be occlusive. A cotton pad is appropriate and can be circular or square, but should approximate 4–6 cm^2. Restraint using an appropriate restrainer is preferred to assure occlusion. If wrapping is used, additional exposures may be required.

Day 0 – control group

- One flank is cleared of hair (closely-clipped). The vehicle only is applied in a similar manner to that used for the treated group. The test patch system is held in contact with the skin by an occlusive patch or chamber and a suitable dressing for 6 hours. If it can be demonstrated that a sham control group is not necessary, a naive control group may be used.

Days 6–8 and 13–15 – treated and control groups

- The same application as on day 0 is carried out on the same test area (cleared of hair if necessary) of the same flank on day 6–8, and again on day 13–15.

Challenge

Day 27–29-treated and control groups

- The untreated flank of treated and control animals is cleared of hair (closely-clipped). An occlusive patch or chamber containing the appropriate amount of test substance is applied, at the maximum non-irritant concentration, to the posterior untreated flank of treated and control animals.
- When relevant, an occlusive patch or chamber with vehicle only is also applied to the anterior untreated flank of both treated and control animals. The patches or chambers are held in contact by a suitable dressing for 6 hours.

Observations-treated and control groups

- Approximately 21 hours after removing the patch the challenge area is cleared of hair;
- Approximately three hours later (approximately 30 hours after application of the challenge patch) the skin reactions are observed and recorded according to the grades shown in the Guinea-Pig Maximization Test (Table 3.17);
- Approximately 24 hours after the 30-hour observation (approximately 54 hours after application of the challenge patch) skin reactions are again observed and recorded.

Blind reading of test and control animals is encouraged.

Table 3.17. Magnusson and Kligman Grading Scale for the evaluation of Challenge Patch Reactions for the Buehler Test Method

0	No visible change
1	Discrete or patchy erythema
2	Moderate and confluent erythema
3	Intense erythema and swelling

Rechallenge

If it is necessary to clarify the results obtained in the first challenge, a second challenge (i.e., a rechallenge), where appropriate with a new control group, should be considered approximately one week after the first one. The rechallenge may also be performed on the original control group.

Clinical observations

All skin reactions and any unusual findings, including systemic reactions, resulting from induction and challenge procedures should be observed and recorded. Other procedures, e.g., histopathological examination, measurement of skin fold thickness, may be carried out to clarify doubtful reactions.

Ref: OECD Test Guideline 406.

3.3. IRRITATION

Chemicals released from medical devices that come into contact with the body may produce skin, mucosal or eye irritation. These are known as local tissue responses which are often characterized by signs of inflammation, redness or swelling. In some instances, these can be accompanied by heat and/or pain. Common examples of medical device-related chemicals which are capable of causing irritation include residual concentrations of ethylene oxide (EtO) and chemical detergents.

Irritation, as defined by the relevant ISO standard, is commonly defined as a "localized inflammatory response to single, repeated, or continuous application of the test substance, without involvement of an immunological mechanism." The standard also provides guidance on how a manufacturer should approach assessment of the potential of a material to cause irritation. A device manufacturer should first conduct a review of any available literature to determine whether other manufacturers have reported that the chemical or material under consideration can cause irritation. The manufacturer should also ensure that accredited and validated *in vitro* tests be used to assess the chemical component of interest. These steps will help manufacturers to rule out potentially harmful substances very early on before proceeding to *in vivo* testing for evaluation. Manufacturers should then use non-invasive clinical studies in human subjects for final confirmation. The need for such clinical trials is usually dependent on the risk class and availability of literature.

Irritation testing is similar to sensitization testing however irritation testing focuses on the ability of a medical device to cause an immediate reaction when applied directly. The irritation response is qualitatively graded using a primary irritation index (PII) that indicates the severity of the reaction from negligible to severe. There are several methods for irritation testing however these are selected and performed using a model appropriate for the route of exposure (skin, eye, mucosa) and duration of exposure or contact. These include the intracutaneous test, the primary skin test and the ocular irritation test; all of which are *in vivo* and non-clinical.

3.4. THE INTRACUTANEOUS TEST

The intracutaneous test is conducted much like allergy testing is done in human patients. Fluid extracts of the test material are prepared under controlled conditions. A known volume of the fluid extract is then injected intracutaneously at multiple sites on the shaved backs of albino rabbits. An equal number of control sites are injected with an untreated volume of the extraction vehicle. The test and control sites are then observed and scored at 24, 48, and 72 hours after injection for the severity of any redness (erythema) or swelling (edema). Extracts that produce a significantly greater response than controls are considered as irritants. This form of testing is often said to be aggressive in that it makes use of extracts prepared under exaggerated conditions and places them directly into the skin of the test animal. In this way, it maximizes the chance of finding chemicals with an irritation potential.

3.4.1. Primary Skin Irritation Test

The primary skin irritation test is less aggressive in that portions of the test material itself are simply placed on the shaved backs of albino rabbits rather than being injected into the animal. The samples are then covered with an occlusive dressing and left in place for between 4–24 hours. The contact sites are then observed for an additional period of up to 72 hours and scored for erythema and edema by comparing obtained results with known values for primary skin irritation.

3.4.2 Ocular Irritation Test

The ocular irritation test is usually performed with fluid extracts although some materials may be tested directly as solids or powders. A small volume of the test extract is placed directly into the pocket formed when pulling the lower eyelid of an albino rabbit. The rabbit's other eye is left untreated as a control. Observations are made at regular intervals for up to 72 hours. The reactions are then scored for redness and swelling of the eye's conjunctiva, the response of the iris to light, corneal opacity, and presence of discharge. These scores are then compared with a classification table to determine which test materials are considered eye irritants.

Box 6: Overview of the Ocular Irritation Test

Ocular irritation test

The premise of ocular irritation is on an assessment of the potential of the material under test to produce ocular irritation. The ocular irritation test should only be carried out if irritation data cannot be obtained by other means, and only for those materials that will come into contact with the eye or eyelid. Those test materials and/or final products that have demonstrated severe irritation in a dermal study should not be tested for eye irritation. In addition, any material where literature or previous testing has identified that they are skin irritants or those with a pH of less than or equal to 2.0 or greater than or equal to 11.5 should not be tested. Instead they should automatically be labeled as a potential eye irritant.

If the test material is a liquid, then 0.1 ml of the undiluted test extract should be inserted into the lower conjunctival sac of one eye. If the test material is a solid or granular-like product, the it should be ground up into a fine dust and an amount occupying a volume of 0.1 ml should be inserted into the into the lower conjunctival sac of one eye. Test materials should be prepared as per guidelines laid out in ISO 10993-10.

Test animals

Healthy young adult albino rabbits of either sex from a single strain, weighing approximately 2 kg to 3 kg, should be used. The test animals should be prepared and acclimatized as well as cared for s per guidelines laid out in ISO 10993-2. One test animal should be used for initial testing. If a positive response is seen then additional testing should be carried out. However, if no response is observed for solid or liquid materials then a minimum of two further animals should be tested. Additional testing can be considered in three animals are used and the response is still not clear.

Test procedure

Visual examinations should be conducted on both eyes of each rabbit for evidence of ocular abnormality. This should be done no longer than 24 hours before starting the ocular irritation test. If either eye shows any abnormality, the rabbit should be replaced. The test sample shall be inserted as per guidelines defined in ISO 10993-10. If repeated exposure to the material is anticipated and the test material has not demonstrated a significant response in the acute test, a repeat-exposure study can be conducted. Repeated exposure shall only be carried out after completion of the acute exposure test for a period of 72 hours. The duration of the exposure should bear resemblance to the length of use of the test material/device in the clinical situation.

Observation of animals

For animals receiving a single treatment, both eyes of each animal shall be examined after 1 hour, 24-hour, 48 hour and 72 hours. Extended observations may be necessary if there are persistent lesions, in order to determine the progress of such lesions or reversal of effects. If severe lesions are seen then extended observations should not be carried out. Reactions will be graded as per the scoring system in ISO 10993-10 for ocular lesions. A summary table of how to grade ocular lesions is provided in table 3.18.

Table 3.18. Numerical grading scale for ocular irritation testing

Part of the Eye	Reaction	Numerical Grading
Cornea	Degree of opacity (most dense area)	
	No opacity	0
	Scattered or diffuse areas, details of iris clearly visible	1*
	Easily discernible translucent areas, details of iris slightly obscured	2*
	Opalescent areas, no details of iris visible, size of pupil barely discernible	3*
	Opaque, detail of iris not visible	4*
	Area of the cornea involved	
	One-quarter (or less), not zero	0
	Greater than one-quarter, but less than half	1
	Greater than half, but less than three-quarters	2
	Greater than three-quarters, up to whole area	3

iris	Normal	0
	Folds above normal, congestion swelling, circumcorneal injection (any or all or combination of these), iris still reacting to light (sluggish reaction is positive)	1*
	No reaction to light, hemorrhage, gross destruction (any or all of these)	2*
conjunctivae	**Redness (refers to palpebral and bulbar conjunctiva excluding cornea and iris)**	
	Vessels normal	0
	Vessels definitely injected above normal	1
	More diffuse, deeper crimson red, individual vessels not easily discernible	2*
	Diffuse beefy red	3*
	chemosis	
	No swelling	0
	Any swelling above normal (including nictitating membrane)	1
	Obvious swelling with partial eversion of lids	2*
	Swelling with lids about half-closed	3*
	Swelling with lids about half-closed to completely closed	4*
	Discharge	
	No discharge	0
	Any amount different from normal (does not include small amounts observed in inner canthus of normal animals)	1
	Discharge with moistening of the lids and hairs just adjacent to lids	2

	Discharge with moistening of lids and hairs, and considerable area around the eye	3

* indicates a positive result.

In cases where multiple treatments of the test material are involved, both eyes of each animal should be examined immediately before and approximately 1 hour after each installation.

If there is evidence of irritation after the last treatment, then observations may be extended. Reactions should be graded and recorded as per guidelines in ISO 10993-10. Animals may be immediately withdrawn and euthanized from the study for various conditions. These include:

- Very severe ocular damage;
- Blood-stained or purulent discharge;
- Significant corneal ulceration; and
- Any animal showing maximum effects on the grading system.

Evaluation of results

Results should be evaluated such that differences between the test eye and control eyes are characterized. For acute exposure, if the treated eye in more than one animal shows a positive result at any of the observations, then the material is considered an eye irritant and further testing is not required. However, if only one of the three treated eyes show a mild or moderate reaction, then further animals should be treated. The test material is considered to be an eye irritant if more than half of the eyes treated in the test group exhibit a positive result at any stage of the observation. A severe reaction in only one animal is considered sufficient to label the material as an eye irritant.

For repeated exposure the test material is considered an eye irritant if more than half of the animals in the test group exhibit a positive result at any stage of the observation.

3.4.3. Other Irritation Tests

Oral, rectal, penile, and vaginal irritation tests are some of the other test methods that can be used as a complement to the above-mentioned testing. These test methods are considered relevant for medical devices intended to be applied to these respective mucosal areas of the body.

3.5. *IN VITRO* IRRITATION TESTING

The rat transcutaneous electrical resistance (TER) test and the Human skin model test are commonly used for *in vitro* irritation testing. These test methods have been validated and accepted as alternative tests to assess the irritation potential of chemicals as per OECD Guidelines 430 and 431. To date, many national and international organizations continue to focus on the research and development of *in vitro* tests for skin irritancy while others have been working to quantify the response of animals and humans in order to better define endpoints using non-invasive techniques.

Box 7: Overview of the *in Vitro* Tests for Skin Irritation Using Human Skin Models

***In vitro* tests for skin irritation**

The main concept of the *in vitro* skin model irritation assay is based on the fact that irritant chemicals are able to penetrate the stratum corneum by diffusion and are cytotoxic to the cells in the underlying layers. Moreover, if the cytotoxic effect is absent or weak, a quantifiable number of inflammatory mediators are released by the epidermis and may be used in a tiered approach to increase the sensitivity of the test. The test material is applied topically to a three-dimensional human epidermal model, comprised of at least a reconstructed epidermis with several epidermal cell layers and a functional stratum corneum. Irritant materials are identified by their ability to decrease cell viability below defined threshold levels such as 50% for example. As an additional measure of skin irritation, release of inflammatory mediators can be determined.

Common examples of human skin models that can be obtained commercially include EpiDerm™, EPISKIN™, Vitrolife-Skin, TESTSKIN, and the Labcyte EPI-MODEL to name a few. Human skin models can also be developed or constructed in a testing laboratory and any new model should be validated and compared with existing models. Ideally, human keratinocytes should be used to construct the epithelium of these skin models and they should represent multiple layers of viable epithelial cells under a functional stratum corneum. The stratum corneum should be multi-layered and should contain the essential lipid profile to produce a functional barrier that can resist the rapid penetration of cytotoxic marker chemicals. The skin model should be free of contamination by bacteria, mycoplasma, or fungi.

Test procedure

The tissue constructs used in human skin models are commonly cultured in tissue culture medium according to protocols provided by the manufacturer. If the tissues have been transported, care should be taken to apply the recommended procedures for conditioning before the use of the tissues in the laboratory. At least three tissue replicates of sufficient size should be incubated with the test samples and controls. The test samples and controls should be incubated with the skin constructs for at least 15 minutes. A sufficient amount of test material should be applied uniformly over the skin surface. Solid substances should be moistened with deionized or distilled water after application to ensure good contact with the skin. Solids should be ground to a powder before application. After the 15-minute incubation period, the test samples are removed by thorough washing and rinsing with an appropriate buffer or 0.9% NaCl. The washing and rinsing procedure should be adequate to remove all test materials. A post-exposure recovery period should be included to allow for recovery from any weak irritant effects. Cellular survival rates are determined after 42 hours.

After (42 ± 2) h, cellular survival is determined. It should also be noted that different medical devices can contain varying concentrations of irritating chemicals and therefore it may be necessary to adapt the exposure period.

Measuring cell viability

Only quantitative methods can be used to measure cell viability. In addition, the measure of viability should be compatible with use in a three-dimensional tissue construct. Non-specific dye binding should be taken into consideration such that it does not interfere with the viability measurement. For this reason, protein binding dyes and those that do not undergo metabolic conversion are not appropriate. Commonly used assays include the MTT reduction assay in which the skin sample is placed in MTT solution of appropriate concentration for 3 hours. The precipitated blue formazan product is then extracted using an isopropanol solvent and the concentration of formazan is measured by determining the OD at a wavelength between 540 nm and 595 nm.

Evaluating results

The percentage viability can be calculated using the optical density (OD) values obtained per test sample when compared to the negative control. The test substance is considered to be irritant to skin if in the EPISKIN™ or EpiDerm™ model the tissue viability after exposure and post-incubation is equal to or less than 50%. In the EpiDerm™ model, a negative response requires confirmation in the rabbit skin irritation test.

Ref: ISO 10993-10: Biological evaluation of medical devices: Tests for irritation and skin sensitization.

3.6. *IN VIVO* IRRITATION TESTING

A range of factors should be taken into consideration when *in vivo* irritation testing is to be conducted. These are listed in the box below. It should be noted that irritation testing of medical devices can be performed with the finished product and/or extracts of the device itself. Guidelines are provided in the ISO 10993-10 standard and whilst flexibility with respect to the actual detail followed is allowed, consistency is important as this contributes to the comparability of test results contained across different material and laboratories.

Box 8: Factors Affecting the Results Obtain in Irritation Studies

- The nature of the device used in a patch test;
- The dose of the test material;
- The method of application of the test material;
- The degree of occlusion;
- The application sites;
- The duration and number of exposures; and
- The techniques used in evaluating the test.

3.6.1 Animal Irritation Testing

This form of testing involves an assessment of the potential of the test material to produce some form of dermal irritation in a relevant animal model, often the rabbit.

3.6.1.1. Preparation of Test Materials

The test material can be tested in either solid or liquid form but each will require different preparation. Solid materials that are a form that is suitable for testing as a solid, such as sheets or films; shall be tested without modification. Samples should be prepared as a 2.5 cm x 2.5 cm size and of a thickness within a 0.5 cm range. The solid may be broken down however care should be taken to ensure no contamination occurs during these processes. Solids can be moistened sufficiently with water or a suitable non-irritant solvent to ensure good contact with the tissues. It should however be remembered that physicochemical properties of some test materials (such as ceramics) can be altered during such processing. Appropriate controls should be prepared in the same way. A solid may also be tested by preparing extracts from the solid however these should be prepared as described in ISO 10993-12. Polar, non-polar, and/or additional solvents should be used and these should be non-irritant. Liquid test samples should be tested in an undiluted form by direct deposition onto the test area. If needed, an appropriate solvent can be used. Sterility of test materials is also important. If the final product or device is supplied in a sterile condition, then the test material should be sterilized using the same process used to sterilize the final product prior to testing. It should be noted that products sterilized by EtO present a technical difficulty as EtO and its reaction products can produce a biological response. For this reason, to enable differentiation between the effects produced by the test material and those produced by EtO residuals, manufacturers should consider testing the device pre-and post-ethylene oxide sterilization.

3.6.1.2. Selection of a Suitable Animal Model

The basis of animal irritation testing is to determine the ability of the test material to produce dermal irritation in a suitable animal model. Rabbits are generally the preferred test animal. In such testing, healthy young adult albino rabbits of either sex from a single strain is used. If irritation is anticipated, consideration shall be given to testing in one animal first. Guidelines for animal husbandry is laid out in ISO 10993-2.

Animal preparation for these tests are important as the condition of the skin forms a critical factor. Only animals with healthy intact skin should be used. The fur on the backs of the test animals is generally clipped within 4–24 hours of testing. This is done a sufficient distance on both sides of the spine and of a suitable size of approximately 10×15 cm^2. management of animal fur is essential to facilitate observation and/or to accommodate repeated exposures.

3.6.1.3. Application of the Test Sample

Depending on the type of test sample (powder, liquid, solid, extract or extract vehicle), different protocols for application will be followed. For a powdered or liquid sample, approximately 0.5 g or 0.5 ml of the test material should be applied directly to each test skin site (Figure 3.2). The site should be slightly moistened before application with water (or other suitable solvents) if the material is a powder. For solids and hydrophobic materials, there is no need for moistening. The application site should be covered appropriately for the period of contact. At the end of the contact time, the dressings should be removed and the position of the application sites marked with permanent ink.

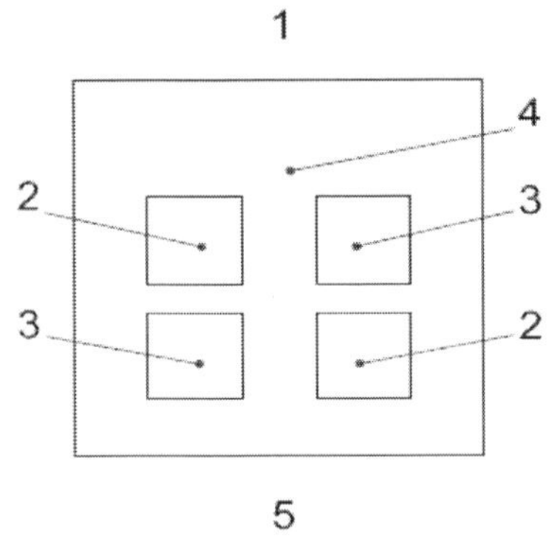

Figure 3.2. Location of application sites on skin of test animal (Key: 1: cranial end, 2: test site, 3: control site, 4: clipped dorsal region, 5: caudal end).

Extracts are usually applied suing absorbent gauze patches. A suitable volume of the test extract should be used such that it allows for sufficient saturation of the gauze. A volume of about 0.5 ml is suitable for a 2.5 cm x 2.5 cm absorbent gauze patch. The patches are applied in a similar manner as the Figure above and with a bandage for a minimum of 4 hours. At the end of the contact time, the dressings are removed and the positions of the sites are marked with permanent ink. Solid test samples are applied directly to the skin on each side of the test rabbit as shown in above figure. When testing solids, the test material shall be moistened sufficiently with water or a suitable solvent to ensure good contact with the skin.

3.6.1.4. Evaluation of Results

For single exposure tests the appearance of each application site should be evaluated and recorded at 1 hour, 24-hour, 48 hour and 72-hour intervals following removal of the patches. Extended observations may be necessary if there are persistent lesions. The primary irritation index (PII) should be determined as follows: only the 24-hour, 48 hour and 72-hour observations should be used for calculations. Observations made prior to dosing or after 72 hours are not used. After the 72-hour grading, all erythema grades plus edema grades at various intervals are totaled separately for each test sample and blank per test animal. The primary irritation score per test animal is calculated by dividing the sum of all the scores by 6 (i.e., two test/observation sites at three time points). To obtain the PII for the test sample, add all the primary irritation scores of the individual animals and divide by the number of animals. This is usually three as three test animals are used per test. When blank or negative controls are used, calculate the primary irritation score for the controls and subtract that score from the score obtained for the test sample to obtain the primary irritation score.

Repeated exposure tests should only be carried out after completion of an acute single-exposure test. The appearance of the application site should be recorded at 1 hour after the removal of the patches and immediately prior to the next application. The number of exposures can vary but after the last exposure, the appearance of each application site should be recorded at 1 hour, 24-hour, 48 hour and 72-hour intervals. Extended observations may be necessary if there are persistent lesions. For repeated exposure assays the primary irritation score for each test animal shall be calculated in the same way as described for single exposure tests. The cumulative irritation index shall then be calculated as follows:

The irritation scores of all test animals are added together and then divided by the total number of animals. This value is known as the cumulative irritation index. This index is then compared with the categories of irritation response defined in Table 3.19.

Table 3.19. Primary or Cumulative Irritation Index Categories

Mean Score	Response Category
0–0.4	Negligible
0.5–1.9	Slight
2–4.9	Moderate
5–8	Severe

The use of natural or full-spectrum lighting should be used when visualizing the skin reactions. These should be described and scored for erythema and edema according to the scoring system outlined in table 3.20.

3.6.1.5 Scoring System for Skin Reactions

After testing is carried out, the irritation response is qualitatively graded using an irritation scale which is similar to the scale used in the M and K Assay. This irritation index scores the severity of irritation reactions from negligible to severe. Skin reactions are generally scored for erythema and edema reactions as per Table 3.20.

Table 3.20. Scoring System for Skin Reactions as per ISO 10993-10: Biological Evaluation of Medical Devices (Tests for Irritation)

Reaction	**Irritation Score**
Erythema and eschar formation	
No erythema	0
Very slight erythema (barely perceptible)	1
Well defined erythema	2
Moderate erythema	3
Severe erythema (beet-red) to eschar formation (preventing grading of erythema)	4
Edema formation	
No edema	0
Wey slight edema (barely perceptible)	1
Well defined edema	2
Moderate edema (raised approximately 1 mm)	3
Sever edema (raised more than 1 mm and extending beyond exposure)	4
Maximum possible score for irritation	8

* Other adverse changes at the skin site shall be recorded and reported.

Box 9: Overview of the Animal Intracutaneous (intradermal) Reactivity Test

Animal intracutaneous (intradermal) reactivity test

The use of the animal intracutaneous (intradermal) reactivity test is advised for those medical devices intended to be used as an implant and involves intradermal injections of the test extract. Any material shown to be a skin, eye or mucosal tissue irritant or material with a pH lower than or equal to 2.0 or higher than or equal to 11.5 shall not be tested intradermally.

Test samples

The test sample shall be an extract prepared as per requirements of ISO 10993-10 depending on it is a solid, liquid or powder. Healthy young adult albino rabbits of either sex from a single strain shall be used. These should not weigh less than 2 kg and shall be acclimatized and cared for as specified in the ISO 10993-2 standard. A minimum of three animals shall initially be used to evaluate the test material. If irritation is expected or anticipated, then one animal should be tested first. A minimum of two additional animals shall be used unless a well-defined positive response with a score greater than 2 for either erythema or edema is seen.

Test procedure

Approximately 4–18 hours before testing, the fur on the backs of the test animals should be clipped such that there is a sufficient distance on both sides of the spine for injection of the test extracts. Intracutaneous injection of 0.2 ml of the test extract obtained with polar or non-polar solvent should be done at five sites on one side of each rabbit (Figure 3.4). Similar injections should be done for the polar or non-polar solvent control on five sites of the contra lateral side of each rabbit.

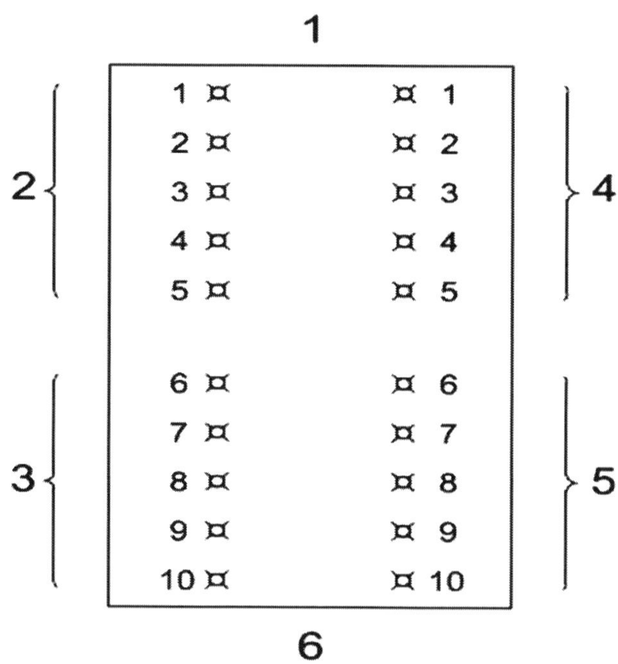

Figure 3.3. Arrangement of injection sites (Key: 1: cranial end, 2: 0.2 ml injections of polar extract, 3: 0.2 ml injections of non-polar extract, 4: 0.2 ml injections of polar solvent control, 5: 0.2 ml injections of non-polar solvent control, 6: caudal end).

Observation of test animals

The appearance of each injection site should be observed and documented immediately after injection and at 24 hours, 48 hours and 72 hours. The tissue reaction should be graded for the presence of erythema and edema according to a defined scoring system as per ISO 10993-10.

Evaluation of results

After grading at 72 hours, all erythema grades plus edema grades at the respective intervals (24 hours, 48 hours and 72 hours) are totaled separately for each test sample or blank for each individual test animal. To calculate the score of a test sample or blank on each individual animal, divide each of the totals by 15 (3 scoring time points with 5 test or blank sample injection sites). The final test sample score can be obtained by subtracting the score of the blank from the test sample score. The requirements of the test are met if the final test sample score is 1.0 or less (Table 3.21)

Table 3.21. Grading System for Intracutaneous (intradermal) Reactions

Reaction	Irritation Score
Erythema and eschar formation	
No erythema	0
Very slight erythema (barely perceptible)	1
Well defined erythema	2
Moderate erythema	3
Severe erythema (beet-red) to eschar formation (preventing grading of erythema)	4
Edema formation	
No edema	0
Wey slight edema (barely perceptible)	1
Well defined edema	2
Moderate edema (raised approximately 1 mm)	3
Sever edema (raised more than 1 mm and extending beyond exposure)	4
Maximum possible score for irritation	8

Other adverse changes at the injection sites shall be recorded and reported.

Ref: ISO 10993-10: Biological evaluation of medical devices: Tests for irritation and skin sensitization

Box 10: Overview of the Human Skin Irritation Test

Human skin irritation test

Currently the prediction of human cutaneous irritation relies primarily on the use of experimental test animals which has led to a few problems when extrapolation from animals to humans. For this reason, chemicals to which known human exposure is high, such as cosmetics and detergents, risk assessments are performed using human skin patch tests. Such human studies can serve several purposes. This includes the 1) direct identification of human hazards by testing chemicals in humans rather than in laboratory animals; 2) provision of risk assessment of certain chemicals to which human exposure is high and 3) facilitation of extrapolation to humans of data obtained previously from laboratory animal studies.

The aim of the human skin irritation test is to determine whether a material presents a significant skin irritation hazard following acute exposure. Clinical tests shall be performed in accordance with ISO 14155-1 and ISO 14155-2. It should be noted that such test materials should not be tested on humans if the following criteria are met:

- They have been shown to be an irritant in *in-vitro* or *in vivo* testing;
- They have been shown to be corrosive in *in-vitro* or *in vivo* testing;
- They present a risk of skin or respiratory tract sensitization;
- They present any acute toxicity hazard under test conditions; and
- They present any genotoxic, reproductive or carcinogenic hazard.

Simplified description of the test method

A single dose of the test material is applied under occlusion to the skin of human volunteers. Irritation is kept to a minimum by applying the test material for short periods. Longer exposure periods can also be appropriate under certain circumstances. Healthy human volunteers should be selected who are at least 18 years of age, not pregnant and not breast-feeding. In addition, human volunteers with a known sensitivity to the test material or showing any signs of dermatitis shall be excluded from the test. The selection of volunteers should also be supervised by a dermatologist or other qualified person.

Liquid test materials are generally used undiluted, however, when testing solids, the test material should be moistened with a small amount of water or where necessary with another suitable vehicle in order to ensure good contact with the skin. The structure of the solid shall be taken into consideration and the choice of test material preparation shall be justified as per ISO 10993-10.

When extract vehicles are used, the influence of the vehicle on irritation of the skin should also be taken into account. If a vehicle other than water is used as the wetting agent for solid compounds, then the application of a blank liquid patch on each subject should also be considered.

At least 30 volunteers should participate and complete the test with no less than one-third of either sex.

Application of the test material

The test material should be applied to intact skin at a suitable site such as the upper outer arm. This should be done by means of an occlusive chamber containing a gauze pad. The patch should measure at least 1.8–2.5 cm and the application site should be the same in all volunteers. The patch shall be held in contact with the skin by means of a suitable non-irritating dressing for the duration of the exposure period. The patch shall deliver an adequate dose of the test material per unit area (approximately 50 mg to 100 mg test material per square centimeter is considered optimal).

Duration of exposure

To avoid unacceptably strong reactions, a cautious approach should be adopted. The patches are applied progressively starting with durations of 15 and 30 minutes, going up to 1 hour, 2-hour, 3 hour and 4 hours. The 15 min and/or 30 min exposure periods may be omitted if there are sufficient indications that excessive reactions will not occur following the 1-hour exposure. Progression to longer exposures, including 24 hours will depend upon the absence of skin irritation. This is important to ensure that any delayed irritant reaction is adequately assessed.

If it is suspected that the material might produce severe irritation, a substantially reduced exposure time shall be employed. This can be done in a pilot group of volunteers. The progress of the study can then be defined on the basis of the data produced.

Clinical observations

Treatment sites are examined for signs of irritation and the responses are scored immediately after patch removal. This is done at 1 hour, 2-hour, 24-hour, 48 hour and 72-hour intervals. The 72-hour interval period is generally included only if necessary, to determine reversibility of the response. The condition of the skin before and after the test shall be described and skin irritation shall be graded and recorded according to a grading scale as defined by ISO 10993-10. A possible grading scale is outlined in Table 3.22.

Table 3.22. Human Skin Irritation Grading Scale

Description of Response	**Grading**
No reaction	0
Weakly positive reaction (usually characterized by mild erythema and/or dryness across most of the treatment site)	1
Moderately positive reaction (usually distinct erythema or dryness, possibly spreading beyond the treatment site)	2
Strongly positive reaction (strong and often spreading erythema with edema and/or eschar formation)	3

For volunteers who have a grading of 1 or greater following an exposure of less than 4 hours, it is assumed that they will present a stronger reaction if exposed to the material for 4 hours. Once a grading of 1 or greater has been obtained, there is no need to subject the reacting volunteer to further treatment with the test material. As with any clinical trial, further observations may be required for proper volunteer care. In addition to irritation, any other effects should be recorded and volunteers should be trained to make related comments while assessors should be trained to note immediate responses when the patches are removed.

Evaluating data

The main focus of the human skin irritation test is to determine whether a material presents a significant skin irritation potential hazard after acute exposure. Therefore, if the test material produces a frequency of skin irritation in the test subjects which is similar to, or greater than, the positive control, it shall be regarded as a significant skin irritant. On the other hand, if it produces a frequency of skin irritation in the test subjects which is substantially and significantly less than the positive control, then it shall not be regarded as a significant skin irritant. Individual variation in the susceptibility to skin irritation should not be confused and should be taken into consideration.

Ref: ISO 10993-10: Biological evaluation of medical devices: Tests for irritation and skin sensitization.

3.7. SYSTEMIC TOXICITY

Systemic toxicity involves evaluating the potential adverse effects of medical devices on organs or tissues that are remote from the site of contact. This can include effects experienced on behavior, general health as well as target organs such as the liver, heart, kidneys, and/or brain. In most cases, tests are performed using appropriate fluid extracts but, in some cases, can also be performed using actual material or product samples. Mice and rats are generally used in systemic toxicity studies however, other species of laboratory animals may also be used for specific products.

The ISO 10993-11 standard identifies systemic toxicity test methods that call for various routes of sample administration and which categorizes adverse effects based on the toxicity effects experienced. These can be acute, sub-acute, subchronic and chronic. Acute effects occur within 24 hours after a single dose or repeated doses are delivered to the test animal while subacute effects appear within 14 to 28 days of delivery. Subchronic effects involve effects seen for up to 10% of an animal's life span while those that continue for longer than 10% of an animal's life span are considered chronic. Depending on the systemic effect being studied, the investigator must select both the route of administration, which should reflect the intended use of the device, and the duration of the study as per ISO 10993-1.

3.7 1. Tests Using Fluid Extracts

The test most commonly used to address acute systemic effects is the systemic toxicity test. This is a relatively simple test that depends upon the visual observations of animals that have received a large dose of extraction fluid prepared under exaggerated conditions of surface area or weight to volume. Saline or oil extracts of the potential device material are prepared and a single dose is administered by intravenous or intraperitoneal injection depending on the type of extract. Generally, intravenous injection is used for saline while intraperitoneal injections are used for oil. These injections are administered to groups of five mice, which are then weighed. Control mice are injected with extraction fluids only. The mice are then observed for the three consecutive days for any adverse signs. These are then compared with the control group in order to make a suitable conclusion.

For those studies lasting longer than three days and which involve repeated doses to an extractor device, or rather constant exposure; may include additional steps such as the collection of terminal blood samples, routine hematology, and serum chemistry tests or collection of representative tissue samples for microscopic examination. Longer-term systemic studies may involve larger groups of animals and additional periodic measurements of body weight and blood tests.

3.7 2. Tests Using Fluid Extracts

While most systemic toxicity tests make use of fluid extracts from device materials, some instances involve the use of device implants for more accurate testing. A common example is those devices constructed of absorbable materials as over time, tissue fluids will act on such materials, causing them to degrade. Such resultant breakdown products may cause adverse effects on various organs. In such implant studies, exaggerated amounts of material should be used whenever possible. In addition, it may be necessary for the investigator to calculate the maximum amount of material that would need to be implanted in a human patient and then implant multiples of that amount in experimental animals. Often such placement is intramuscular, subcutaneous or intraperitoneal.

3.7.3. Tests Using Pyrogens

Some compounds, better known as pyrogens which are commonly derived from chemical or bacterial sources are capable of causing a fever when present in the body at sufficiently high doses. These are commonly detected using the rabbit pyrogen test. This test consists of preparing fluid extracts of the test materials, administering the extracts intravenously to test rabbits and monitoring the animals' rectal temperatures over the course of several hours. If a significant rise in temperature is seen, this indicates the presence of pyrogens.

These types of tests often help manufacturers narrow down the source of toxin/contamination as well. For example, the *Limulus* amoebocyte lysate (LAL) reagent test is used to detect bacterial endotoxins arising from gram-negative bacteria, on production lots. One of the most common sources of such contaminants is the water used in the manufacturing process. Thus, it should be noted that the presence of such bacterial endotoxins on finished products may indicate that while the device itself is biocompatible, there is, however a manufacturing control issue.

3.7.4. Selecting Animal Species

While there are no absolute criteria for selecting a particular animal species for systemic toxicity testing, it should be remembered that the species used should be scientifically justified and in line with the requirements of ISO 10993-2. For most acute oral, intravenous, dermal, and inhalation studies the mouse or rat is preferred. In the case of dermal and implantation studies, the rabbit is most often used. In addition, a single animal species and/or strain should be used when a series of systemic toxicity studies are being done in order to control the variability between species and strains that may occur.

When conducting such studies, healthy young adult animals of known origin and with defined microbiological health status that have been specifically bred for testing should be used. The weight of the animals should be monitored and should not exceed ± 20% of the mean weight per sex of the animal. If females are being used, they should not be pregnant. The requirements of ISO 10993-11 should be met with regards to animal husbandry. This means that animals should be allowed to acclimatize to the laboratory conditions prior to treatment. Any dietary constituents and bedding materials that may produce or influence toxicity should be properly characterized and their potential to influence the outcomes of testing should be taken into consideration.

Size of groups of animals tested is largely dependent on the dose level. Ideally animal group sizes should increase with the duration of treatment. Recommended minimum group sizes are presented in Table 3.23.

Table 3.23. Recommended Minimum Group Sizes

Type of Testing	Rodent	Non-Rodent
Acute	5	3
Sub-acute	10 (5 per sex)[1]	6 (3 per sex)[1]
Subchronic	20 (10 per sex)[1]	8 (4 per sex)[1]
Chronic	40 (20 per sex) [2, 3]	[3]

[1] Testing in a single sex is acceptable. When a device is intended for use in only one sex, testing should be done in that sex.

[2] One dose-level group testing. Where additional exaggerated dose groups are included, the recommended group size may be reduced to 10 per sex.

[3] Expert statistical; consultation for chronic study group size is recommended. The number of animals tested should be based on the minimum required to provide meaningful data. Enough animals must remain at the termination of the study to ensure proper statistical evaluation of the results.

3.7.5. Routes of Exposure/Administration

Test extracts may be administered into the body by multiple routes. In each instance, the route of exposure should be the most clinically relevant to the use of the device. In some instances, an alternative route of exposure may need to be carried out. In these cases, this is acceptable and must be properly justified. Commonly accepted routes of exposure are outlined in Table 3.24.

Table 3.24. Commonly Accepted Routes of Exposure When Carrying Systemic Toxicity Testing

Route of Exposure	Simplified Explanation
Dermal	• Suitable for surface devices

implantation	• Suitable for implanted devices. • May be appropriate for direct testing of material by application to a general or specified area. • The shape and texture of the test article should be taken into consideration.
Inhalation	• Suitable for devices with a contact environment that may occur via inhalation.
intradermal	• Suitable for devices with an intradermal contact environment conducive to chemical leaching. • Test samples are usually administered directly to the intradermal region by injection. • Multiple treatment sites are possible and should be clearly described and justified.
intramuscular	• Suitable for devices with a muscle tissue contact environment conducive to chemical leaching. • Test samples are administered directly to the the • muscle tissue by injection or surgical implantation. • Sites should be chosen to minimize the loss of function or • the possibility of pain from nerve damage as a result of the injected or implanted test sample. • Sites should be rotated for repeated dose studies • Multiple treatment sites are possible and should be clearly described and justified.

Intraperitoneal	• Suitable for devices with a fluid-path or peritoneal cavity contact environment conducive to chemical leaching • Appropriate route when the test extract should not be administered intravenously (for example with non-polar oil extracts or extracts with particulates present) • Test samples are typically administered directly to the peritoneal cavity. • Frequency of doses administered should take into consideration that test extracts by this route are absorbed primarily through the portal circulation and therefore must pass through the liver before reaching general circulation. • Care should be taken not to inject into the stomach or intestinal tract.
intravenous	• Suitable for devices with a direct or indirect fluid-path or blood contact environment conducive to chemical leaching • Test extracts are typically placed in or administered directly to the vascular system • Sample filtration or administration via the intraperitoneal route should be considered if particles are present • Care should be taken to minimize the possibility of extra vascular injection of test extracts • In cases where injecting will exceed 5 minutes, consideration should be given to the use of a butterfly needle or an intravenous cannula.

Oral	• Suitable for devices contacting the oral mucosa directly or indirectly, or for products with other enteral application • Test samples are typically administered by gavage • Experimental animals should be fasted prior to administration of the test extract • Animals should be weighed and then the test sample administered in a single dose based on body weight • After the test sample has been administered, food may be withheld for an additional 3 h to 4 h • in cases where a dose is administered in fractions over a certain period, it may be necessary to provide the • animals with food and water depending on the length of the period.
Subcutaneous	• Suitable for devices with a subcutaneous contact environment conducive to chemical leaching • Test samples are typically administered directly to the subcutaneous region by injection or by implantation • The use of multiple treatment sites is possible and should be clearly described and justified.

3.7.6. Dosing of Test Extracts

Much detail should be paid to the dosing of test extracts in systemic toxicity. Procedures should be designed to avoid any physiological changes or related problems that are not directly related to the toxicity of the test extract/material itself. In cases where a single daily dose of a sufficient volume or concentration cannot be administered suitably, then the dose may be given in smaller fractions over a period not exceeding 24 hours. The test extracts should be delivered at a physiological acceptable temperature and any deviations related should be suitably justified. Suitable guidance on dosage volumes are provided in ISO 10993-11 and summarized in the table below. Justifications should be provided in cases where large dosage volumes are intended to be administered that deviates from that stated in relevant standards. Such large dosage volumes intended for administration via the oral route should be avoided due to overloading of the stomach compacity and reflux into the esophagus (Table 3.25).

Table 3.25. Maximum Dosage Volumes for Test Sample Administration

Species	Subcutaneous (ml/kg)	Intramuscular (ml/kg)	Intraperitoneal (ml/kg)	Gavage (ml/kg)	Intravenous (ml/kg)
Rat	20	1	20	50	40
Mouse	50	2	50	50	50
Rabbit	10	1	20	20	10
Dog	2	1	20	20	10
Monkey	5	1	20	15	10

NOTE: In some instances, the specific regulatory requirements of individual countries may supersede the maximum volumes listed as per standard requirements.

The dosage frequency should be based on clinical relevancy. In acute systemic toxicity studies, the animals should be exposed to the test sample in a single dose or with multiple fractions of the dose given within a 24-hour period. In repeated exposure studies the animals should be dosed with the test sample daily, seven days each week for the duration of the testing.

3.7.7. Clinical Observations

Clinical observations should ideally be performed by trained individuals to ensure for consistent reporting. The frequency and duration of observation should be determined by the nature and severity of the toxic reactions, rate of onset and recovery period. Often an increased frequency of observation may be necessary in the early phase of a study and more so in acute studies. The time at which signs of toxicity appear and disappear, their duration and the time of death are important as well. Observations shall be recorded systematically and records shall be maintained for each animal. Common clinical signs and observations are presented in Table 3.26.

Table 3.26. Common Clinical Signs and Observations

Clinical Observation	Observed Sign	Involved System(s)
Respiratory	Dyspnea (abdominal breathing, gasping) Apnoea Cyanosis Tachypnea Nostril discharges	CNS Pulmonary Cardiac
Motor activities	Decrease/increase somnolence Loss of righting Anesthesia Catalepsy Ataxia Unusual locomotion Prostration Tremors Fasciculation	CNS Somatomotor Sensory Neuromuscular Autonomic
Convulsion	Clonic Tonic Tonic-clonic Asphyxial Opisthotonos	CNS Neuromuscular Autonomic Respiratory
Reflexes	Corneal Righting Myotact Light Startle reflex	CNS Sensory Autonomic Neuromuscular
Ocular signs	Lacrimation Miosis Mydriasis Exophthalmos Ptosis Opacity Iritis Conjunctivitis Chromodacryorrhea Relaxation of nictitating membrane	Autonomic Irritation
Cardiovascular signs	Bradycardia Tachycardia Arrhythmia Vasodilation Vasoconstriction	CNS Autonomic Cardiac Pulmonary

salivation	Excessive	Autonomic
Piloerection	Excessive	Autonomic
Analgesia	Decrease reaction	CNS Sensory
Muscle tone	Hypotonia Hypertonia	Autonomic
Gastrointestinal	Soft stool Diarrhea Emesis Diuresis Rhinorrhea	CNS Autonomic Sensory GI motility kidney
Skin	Edema Erythema	Tissue damage Irritation

3.7.8. Clinical Pathology

Hematology and clinical chemistry analyses are performed in order to investigate the toxic effects in tissues, organs, and other systems of the test animal. These analyses may also be performed on blood samples obtained from repeated exposure study animals. Fasting of animals prior to blood sampling may be necessary in some cases. Urinalysis may be performed during the last week of a long-term repeated exposure study using timed urine volume collection. Common hematology, clinical chemistry and urinalysis parameters for evaluation are listed in Table 3.27.

Table 3.27. Parameters Measured During Clinical Pathology Testing

Clinical Pathology	Parameter Measured
Hematology	• Clotting potential (PT, APTT) • Hemoglobin concentration • Hematocrit • Platelet count • Red blood cell count • White blood cell count • WBC differential

Clinical chemistry	• Albumin • ALP • ALT • AST • Calcium • Chloride • Cholesterol • Creatinine • GGT • Glucose • Inorganic phosphorus • Potassium • Sodium • Total bilirubin • Total protein • Triglycerides • Urea nitrogen • Additional enzymes, as scientifically appropriate • Total immunoglobulin levels may be considered as an indicator for immuno-toxicity
urinalysis	• Appearance • Bilirubin • Glucose • Ketones • Occult Blood • Protein • Sediment • Specific gravity or osmolality • Volume • Other scientifically appropriate tests if test article is suspected to cause specific organ toxicity (generally • requires refrigerated sample collection)

3.7.9. Anatomic Pathology

Gross pathological evaluations should be considered for acute systemic toxicity studies. All animals in repeated exposure studies shall be subjected to a full, detailed gross necropsy which includes careful examination of the external surface of the body, all orifices, and the cranial, thoracic, and abdominal cavities and their contents. Selected organs for weighing should be trimmed of any adherent tissue, as appropriate, and their wet weight taken as soon as possible to avoid drying.

A suggested organ list that should be weighted and preserved in an appropriate fixation medium for histopathological evaluation is presented in the box below.

Box 11: Suggested Organ list for Histopathological Evaluation

- Adrenals*
- All gross lesions (including treatment sites)
- Aorta
- Bone marrow (femur, rib or sternum)
- Brain* (representative sections including cerebrum, cerebellum, and pons)
- Cecum
- Colon
- Duodenum
- Epididymides*
- Esophagus
- Eyes
- Gall bladder (if present)
- Heart*
- Ileum
- Jejunum
- Kidneys*
- Liver*
- Lungs and bronchi (preserved by inflation with fixative and then immersion)
- Lymph nodes (local to cover site of administration and distant to cover systemic effects)
- Mammary gland (female) – Muscle (skeletal)
- Nasal turbinates (for inhalation studies)
- Nerve (sciatic or tibial) preferably in close proximity to the muscle
- Ovaries*
- Pancreas
- Parathyroid
- Pituitary
- Prostate
- Rectum
- Salivary glands
- Seminal vesicles
- Skin
- Spinal cord
- Spleen*
- Sternum

- Stomach
- Testes*
- Thymus*
- Thyroid
- Trachea
- Urinary bladder
- Uterus* (including cervix and oviducts)
- Vagina

* Organs/tissues marked with an asterisk should be weighed with other organs if scientifically appropriate.

The minimum observations for each type of systemic toxicity testing is shown in Table 3.28.

Table 3.28. Summary of Observations

Observation	**Acute**	**Subacute**	**Subchronic/Chronic**
Body weight change	+	+	+
Clinical observations	+	+	+
Clinical pathology	b	a, b	+
Gross pathology	b	+	+
Organ weights	b	+	+
histopathology	b	a, b	+

[a] Chronic systemic toxicity testing is generally a time extension of subchronic testing, justified by the human exposure period. Many of the same parameters are recorded and reported. Group sizes may be increased to include satellite groups for which some, or all, of these observations may be made.

[b] *Consideration should be given to these measurements when clinically indicated or if longer exposure testing is not anticipated.*

Box 12: Overview of Acute Systemic Toxicity Testing

Acute systemic toxicity testing

This type of testing provides an indication on the type of health hazards likely to arise from acute exposure. After administration of the test extract, observations are made and adverse clinical signs, body weight changes, etc., are recorded.

Study design

- Healthy young adult animals are acclimatized to the laboratory conditions for at least 5 days prior to testing;
- Shorter durations shall be justified; and
- Animals are then randomized and assigned to the treatment groups.

Experimental animals

- A rodent species such as a rat or mouse will be used;
- Characteristics of the type of rodent species, number, and sex as well as housing and feeding conditions are described further in ISO 10993-11; and
- If non-rodent species are used, then their use shall be scientifically justified.

Testing conditions

- Appropriate dose levels are described in ISO 10993-11.
- Animals in the control should be handled in an identical manner to the test group subjects.

Procedure

- The test animals shall receive a single dose of the test sample/extract or, when necessary, multiple doses within a single 24-hour period.
- Signs of toxicity should be recorded as they are observed (details should include time of onset, degree, and duration)
- Regular observation of the animals should be carried out
- At the end of the study all surviving animals are euthanized.
- Body weight measurements should be made immediately before dosing, daily for the first three days after dosing, weekly after the first dose if indicated by study duration, and at the end of the study.
- The observation period for an acute systemic toxicity study shall be at least 3 days, or longer when deemed appropriate.
- Clinical pathology evaluations shall be considered (hematology and clinical biochemical determination on blood should be considered)

- Gross pathology evaluations shall be considered. This should include an examination of the external surface of the body, all orifices, and the cranial, thoracic, and abdominal cavities and their contents. When appropriate, consideration should also be given to recording the weight of the brain, liver, kidneys, adrenals, and testes, which should be weighed wet as soon as possible after dissection to avoid drying and subsequent falsely low values.
- Full histopathology is not typically carried out on organs and tissues from animals in the acute systemic toxicity study, unless indicated.

Evaluation of results

- Results obtained should be evaluated in conjunction with the findings of preceding studies, if available.
- These results should be considered in terms of the toxic effects and the gross necropsy findings, if observed.
- The evaluation shall include the relationship between the dose of the test substance and the presence or absence and the incidence and severity of abnormalities, including behavioral and clinical abnormalities, gross lesions, body weight changes, effects on mortality and any other general or specific effects.

Ref: *ISO 10993-11: Biological evaluation of medical devices: Tests for acute systemic toxicity*

Box 13: Overview of Repeated Exposure Systemic Toxicity Testing

Repeated exposure systemic toxicity testing

A more common form of human exposure to many medical devices is in the form of repeated or continuous exposures. Effects from repeated or continuous exposure may potentially occur due to accumulation of chemicals in tissues or by other mechanisms. Repeated exposure systemic toxicity tests provide information on health hazards likely to arise from a prolonged exposure of the test extract.

Study design

- Healthy young adult animals are acclimatized to the laboratory conditions for at least 5 days prior to testing.
- Shorter durations shall be justified.
- Animals are then randomized and assigned to the treatment groups.

Experimental animals

- A rodent species such as a rat or mouse will be used.
- Characteristics of the type of rodent species, number, and sex as well as housing and feeding conditions are described further in ISO 10993-11.
- If non-rodent species are used, then their use shall be scientifically justified.

Testing conditions

- Appropriate dose levels are described in ISO 10993-11.
- Animals in the control should be handled in an identical manner to the test group subjects.
- For longer duration studies, efforts should be made to include at least three dose levels and appropriate controls.

Procedure

- The animals are dosed with the test extract/sample ideally on 7 days/week, over the duration of the study.
- For longer term repeated exposure studies, dosing on a 5 day/week basis is acceptable but should be documented and justified.
- Body weight measurements should be made immediately before dosing, weekly after the first dose if indicated by study duration, and at the end of the study.
- Clinical pathology evaluations shall be considered
- Hematology should be investigated at the end of the test period and more frequent sampling should be considered
- Clinical biochemical determination on blood should be carried out at the end of the test period. More frequent sampling should be considered
- All animals should be subjected to full gross necropsy which includes examination of the external surface of the body, all orifices, and the cranial, thoracic, and abdominal cavities and their contents. The adrenals, brain, heart, kidneys, liver, ovaries, spleen, testes, thymus, and uterus should be weighed wet as soon as possible after dissection to avoid drying and subsequent falsely low values.
- Full histopathology is not typically carried out on organs and tissues from animals in the control and high dose groups

Evaluation of results

- The findings of repeated exposure studies should be evaluated in conjunction with the findings of preceding studies
- The evaluation shall include the relationship between the dose of the test substance and the presence or absence, the incidence and severity, of abnormalities, including behavioral and clinical abnormalities, gross lesions, microscopic changes, identified in target organs, effects on mortality and any other general or specific effects.

3.8. GENOTOXICITY

While most biocompatibility testing focuses on irritation, sensitization, and systemic toxicity testing, there is growing concern that devices and their related components or material extracts may also exert genotoxic effects. For this important reason, any attempt to assess the safety of a high-risk medical device that is intended for intimate body contact or permanent implantation should include testing for the presence of toxins that may exert an effect on the genetic material of cells. Such testing needs are more outlined in greater detail in the ISO 10993 Part 3 standard which details tests for Genotoxicity, Carcinogenicity, and reproductive toxicity.

Any form of alteration in the DNA structure that results in permanent inheritable changes in cellular function is called a mutation. Those agents responsible for such mutations are known as genotoxic agents or genotoxins. There are three major types of genotoxic effects. These are gene mutations, chromosomal aberrations and DNA effects. While there are no single *in vitro* assays capable of detecting all three types of genotoxic effects in one, there are a range of individual tests that can be done gene mutation and chromosomal aberration tests detect actual lesions in the DNA molecule while DNA effects tests can detect events that may lead to cellular damage.

It is important to remember that while many of the ISO 10993 series standards refer to the OECD guidelines for specific test methods. Many biomaterials are insoluble and the OECE methods are designed to test soluble chemicals so for this reason the tests must be modified to accommodate the evaluation of fluid extracts. The United States Pharmacopeia (USP) has established standard preparation methods for materials testing that can be used for genotoxicity testing. This together with ISO 10993-12 (Sample preparation and reference materials) aid in describing standard methods for preparation of extracts of device materials.

3.8.1. Gene Mutation Tests

Mutations affecting a small portion of the DNA molecule including frameshifts and base-pair substitutions are referred to as point mutations. One of the most common test methods used to detect gene mutations in the Ames bacterial reverse mutation assay. This assay utilizes bacteria to test whether a given chemical can cause mutations in the DNA of the test organism. Several strains of histidine dependent *Salmonella typhimurium* are used. These strains are said to be auxotrophic mutants-meaning they require histidine for growth but themselves cannot produce it. If a test substance is thought to be genotoxic, it will create a mutation that will result in the bacterial strains returning to a prototrophic state such that the cells can grow on a histidine-free medium.

The assay involves S9 active rat liver microsomes which are incorporated into a portion of the test organisms to simulate whole-animal exposure.

Following exposure to the fluid extract from the test article, the organisms are plated in triplicate onto histidine-free growth nutrient agar and incubated for a specified period. The colonies are then enumerated and these data are compared to counts obtained for negative and positive control conditions. The unreverted test strains will not grow without histidine which means any further growth indicates that exposure to a genotoxic agent has caused point mutations producing bacterial strains that no longer require histidine (Figure 3.4).

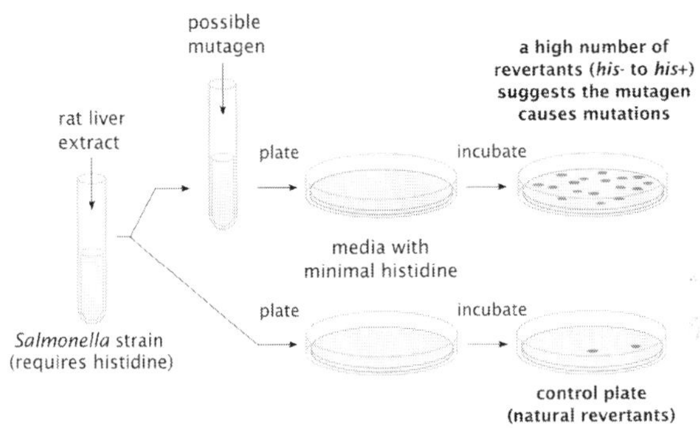

Figure 3.4. Simplified diagram of the Ames test procedure.

Box 14: Simplified Description of The Ames Test

Objective

To determine the mutagenic activity of chemicals by observing whether they cause mutations in sample bacteria.

Method

- Isolate an auxotrophic strain of *Salmonella Typhimurium* for histidine (i.e., Histidine negative)
- Prepare a test suspension of histidine negative *Salmonella Typhimurium* in a plain buffer with the test chemical. Also, add a small amount of histidine. This is often required so the bacterial strain can start growing. Once histidine is depleted only those bacteria that have mutated to gain the ability to synthesize histidine will form colonies.
- Prepare a control suspension of histidine negative S*almonella Typhimurium* but without test chemicals.
- Incubate the suspensions at 37°C for 20 minutes
- Prepare two lines of agar plates-one for the control and one for the test suspension
- Spread the suspension on each line of agar plates
- Incubate the plates at 37°C for 48 hours
- After 48 hours count the number of colonies on each plate.

Result Interpretation

- The mutagenicity of chemicals is proportional to the number of colonies observed.
- If there is a large number of colonies on the test plate in comparison to control, then such chemical is said to be mutagens.
- It should also be noted that very few numbers of colonies can be seen on the control plate. This may be due to spontaneous point mutations on the histidine encoding gene.

3.8.2. Mouse Lymphoma Assay

Another *in vitro* method for detecting gene mutations in a mammalian system is the mouse lymphoma assay. This test uses the L5178Y mouse lymphoma TK+/- cell line to detect mutagenic events by measuring resistance to the lethal nucleoside analog triflurothymidine (TFT). Following incubation, the cultures are cloned in restrictive media for the mutant phenotype. When added to selective medium containing TFT, the wild type TK (+/-) cells dies while those having undergone mutations survive. Mutations are measured at the thymidine kinase locus to detect base-pair mutations, frameshift mutations, and small deletions (Figure 3.3).

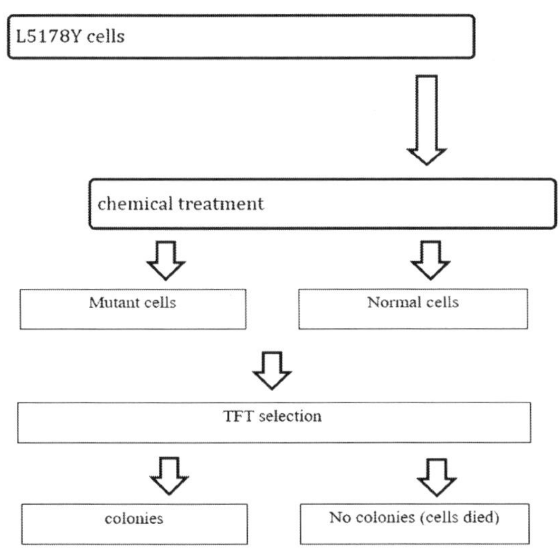

Figure 3.3. Simplified principle of the mouse lymphoma assay.

3.9. IMPLANTATION

Implant studies are generally used to evaluate the biocompatibility of medical devices or related biomaterials that come into direct contact with living tissues other than skin components. Those devices having prolonged or permanent contact with tissues, bones, or dentin and prolonged contact with blood will require testing by implantation. Common examples include sutures, surgical ligating clips and various other implantable devices.

Implantation studies evaluate devices which are intended to be implanted for either short-term or long-term periods. These studies are intended to evaluate both absorbable and non-absorbable materials as well as solid, non-solid, and degradable materials. The ISO 10993-6 Standard: Test for local effects after implantation focuses on assessment of the local effects of an implanted material on living tissue at the macroscopic and microscopic level. Both short term effects and chronic effects are addressed. Local effects are evaluated by comparison of the tissue response caused from a test sample to that caused by control substances used in medical devices whose clinical acceptability and biocompatibility characteristics have already been defined.

Generally, these test methods allow for a minimum implantation of 1 week to allow for normal wound healing around the specimens. After acute wound healing, the tissue response more accurately reflects the steady-state of biological response to the implanted material. The test materials for implantation may be solid on non-solid samples which tested on various animal species as defined by the ISO standard. These include mice, rats, guinea pigs, rabbits, dogs, sheep, goats, or pigs. Selection of the appropriate animal species is based on the type of product as well as the size and shape. Macroscopic evaluation is usually on a cross-section through the middle of the implant specimen. Tissues retrieved for microscopic evaluation may include or exclude the implant specimen depending on the physical characteristics of the material and whether the implant-tissue interface is to be examined. Removing the implant to examine only the tissue envelope destroys the tissue layers closest to the implant specimen. Examination of the tissue-implant interface requires embedding the tissue with the implant in situ using plastic embedding media.

The dynamics of biochemical exchange and cellular and immunologic responses may also be assessed in implantation studies especially through the use of histopathology. Histological assessment includes assessing the extent of fibrosis, fibrous capsule formation and inflammation; tissue necrosis and degeneration; the number and distribution of inflammatory cells and other observations including material debris, granuloma, and tissue in growth. Commonly defined implantation studies include implantation in subcutaneous tissue, implantation in muscle, implantation bone and implantation in brain tissue.

3.9.1. Sample Preparation

Test samples and control as well as reference materials (RM) should be prepared in accordance with the guidelines from ISO 10993-12. Depending on the type of implant and implantation study being carried out, the implant size and shape should be prepared and documented accordingly. It is important to note that the physical characteristics of an implant such as form, density, hardness, and surface can all influence the tissue response. When conducting the testing, implants should be handled aseptically and in such a way as to ensure they are not damaged or contaminated prior to or during implantation. For those devices consisting of two or more different materials, the test articles should be of similar composition or multiple implants may be needed. The test sample should be implanted into the tissues most relevant to the intended clinical use identified. Implantation sites should also be marked in a manner suitable for identification of the site at the end of the designated time periods.

3.9.2. Selecting a Suitable Test Animal Species

The guidelines laid out in ISO 10993-2 should be followed with regards to animal care and accommodation. While small laboratory animals are preferred, the use of larger animals may be justified depending on the particular biomaterial under investigation. Various factors should be taken into consideration such as the size of the implant test samples, the number of implants per animal, the intended duration of the test in relation to the expected lifespan of the animals, as well as how biological responses differ with different species. For short-term testing, animals such as rodents or rabbits are commonly used while rodents, rabbits, dogs, sheep, goats, pigs, and other animals with a relatively long-life expectancy are suitable for long term testing. In some instances, a pilot study in rodents may be considered to determine the expected rate of degradation before embarking on studies on larger animals. The number and size of implants inserted into an animal will depend on the size of the species and the anatomical location and the reference control and test samples should be implanted into the same animal.

3.9.3. Defining the Test Period

The period of testing should be determined based on the defined intended use of the medical device. For non-absorbable materials, short-term responses are normally assessed between 1–4 weeks while the long-term responses are tested for greater than 12 weeks. The local biological response to the test material will depend on both, the properties of the test material itself and on the response to the associated trauma of surgery. For this reason, a period of about two weeks should be factored in after implantation in order to distinguish between tissue reactions evoked from surgery and by the implant itself. For absorbable materials, the test period should be related to the estimated degradation time of the test product at a clinically relevant implantation site. Ideally, the study duration should extend up to or beyond the point of complete absorption. Generally accepted observation periods for non-absorbable biomaterials are given in Table 3.29.

Table 3.29. Possible test Periods for Long-Term Implantation of Biomaterials as Defined by ISO 10993-10: Biological Evaluation of Medical Devices (Tests for irritation and sensitization)

Species	**Implantation Period in Weeks***				
	13	**26**	**52**	**78**	**104**
Mice	X	X	X	-	-
Rats	X	X	X	-	-
Guinea pigs	X	X	X	-	-
Rabbits	X	X	X	X	X
Dogs	X	X	X	X	X
Sheep	X	X	X	X	X
Goats	X	X	X	X	X
Pigs	X	X	X	X	X

* These are commonly used implantation periods however other time frames may be applicable based on specific characteristics of the test material.

3.9.4. Testing Conditions and Evaluation

Depending on the specific type of test being done, the number of implants will vary per animal. These are further described in ISO 10993-6. It is important to note that the surgical technique may profoundly influence the result of implantation. The surgery should be carried out under aseptic conditions and in a manner that minimizes trauma at the implant site to the test animal. Remove the hair from the surgical area should be done by clipping, shaving, or any other mechanical means. The exposed skin should then be disinfected with an appropriate antiseptic. The wound site should be closed using either sutures or wound clips. The health of the test animals should be observed and recorded at regular intervals throughout the study. Following surgery, each animal should be observed at appropriate intervals during the test period, and any abnormal findings should be recorded. The biological response should be evaluated by reviewing the macroscopic and histopathological responses over time. The responses from the test sample should be compared to that obtained from the control sample and graded accordingly. Grading systems have been defined further in ISO 10993-6.

3.9.5. Macroscopic Assessment

Each implant site should be examined for alterations of the normal structure. Use of a lens with low magnification is recommended and the nature and extent of any tissue reaction should be observed and recorded. Common observations include hematoma, edema, encapsulation, and so on. Color macro photography can be used. In addition to the inspection of the implant site, whenever an animal has shown signs of ill health or reactions to the implant, gross necropsy shall be conducted.

3.9.6. Microscopic Assessment

The scoring system used for the histological evaluation should take into account the extent of the area affected. Biological response parameters should be reviewed and recorded. This includes:

- The extent of fibrosis/fibrous capsule inflammation;
- Changes/degeneration of tissue (determined by examining tissue morphology);
- The number and distribution of inflammatory cell types (this includes polymorphonuclear cells, lymphocytes, plasma cells, eosinophils, macrophages, and multinucleated cells);
- The presence and extent of necrosis;
- Other tissue alterations (such as vascularization, fatty infiltration, granuloma formation, mineralization, and bone formation);
- Test material parameters (such as fragmentation and/or debris presence, form, and location of remnants of degraded material);
- The quality and quantity of tissue in growth.

- Histological responses should also be documented. Often photomicrographs are used.

Box 15: Overview of Implantation Testing in Subcutaneous Tissue

Implantation in subcutaneous tissues

This particular test method can be used to determine the biological response of subcutaneous tissue to an implanted material. The test can be used to compare the effect of different surface textures or conditions of the same material, or to assess the effect of various treatments or modifications of a material.

Test samples

It is important to note that implant sizes are often based on the size of the test animal. The following shall be considered for minimum dimensions:

- When using discs, test samples of 10 mm to 12 mm in diameter and from 0.3 mm to 1.0 mm in thickness should be prepared;
- When using rods and cylinders, test samples of 1.5 mm to 2 mm in diameter and 5 mm to 10 mm in length with rounded ends should be prepared;
- Samples that are not solids, such as powders, should be prepared in tubes of 1.5 mm in diameter and 5 mm in length. These materials can be implanted directly into the tissues. A location marker is recommended for absorbable materials; and
- Test samples of other dimensions that are anatomically compatible may also be used.

Test animals and implant sites

Implants should be inserted in the dorsal subcutaneous tissue of adult mice, rats, guinea-pigs, or rabbits. The testing laboratory should select one species in accordance with the provisions of ISO 10993-2. At least three animals should be used per test material. Ideally, there should be a total of 10 tests and 10 control samples for each material and implantation period. When multiple tissue samples are taken from a single implant site, sections for histology shall be at least 1 cm apart.

Implantation procedure and evaluation

Implantation can occur alongside the dorsal midline or in the neck. If implantation is occurring alongside the dorsal midline, an incision should be made in the skin or one or more subcutaneous pockets made by blunt dissection. The base of the pocket should be more than 10 mm from the line of incision. One implant should be placed in each pocket. The implants should not be able to touch one another. In implantation is occurring in the neck, a 10 mm long incision should be made in mice, above the sacrum. A subcutaneous tunnel should be prepared by blunt dissection towards the neck. One implant should then be pushed through the tunnel to position it at the neck. In rats, one implant of each of the control and test materials should be inserted separately on each side of the neck. The implants should be placed such that they do not touch one another. The tunnel should be closed with stitches of appropriate suture material at a certain distance from the implant site so as to prevent the implant from moving. A suitable implantation period should be chosen as defined in ISO 10993-6. Evaluation of results are also conducted as defined in ISO 10993-6.

Box 16: Overview of Implantation Testing in Muscles

Implantation in muscle

This test method is generally used for assessing the biological response of muscle tissue to an implanted test material. The implant is inserted into the muscles of test animals and the biological response to of the test implant sample is compared with the biological response to control implant samples.

Test samples

The preparation of test and control samples are described in ISO 10993-6. It is important to note that implant sizes are based on the size of the muscle group chosen for testing. For rabbit paravertebral muscles, implants of a width of 1 mm to 3 mm and a length of approximately 10 mm are used. Larger samples of up to 10 mm in diameter and 3 mm in thickness may be surgically implanted as well. In addition, implants of other dimensions that are anatomically compatible may be used.

Test animals and implant sites

It is important to ensure that the muscles are of sufficient size to accommodate the implant samples before testing. Only one animal species per test should be used. The implants should be inserted into the muscle of the animals under anesthesia. At least three test animals should be used such that a total of 10 test samples and 10 control samples for each implantation period are carried out. A non-absorbable control sample shall be evaluated at each time point.

Implantation procedure and evaluation

The implantation shall be conducted by hypodermic needles or trocars. For larger implants, other appropriate surgical implantation techniques may be used. Test samples should be implanted into the body of the muscle with the long axis parallel to the muscle fibers. For rabbit paravertebral muscles, sufficient samples of the test materials should be implanted along one side of the spine. This should be approximately 25 mm to 50 mm from the midline and parallel to the spinal column, and about 25 mm apart from each other. Control material should be implanted in a similar way in the contra lateral muscle of each test animal. The implantation period should be determined as per ISO 10993-6 and evaluation of responses shall be observed and documented as per ISO 10993-6.

Box 17: Overview of Implantation Testing in Bone

Implantation in bone

This test method is used for determining the biological response of bone tissue to an implanted material. This type of testing may be used to compare the effect of different surface textures or conditions of the same material, or to assess the effect of various treatments or modifications of a material. The implantation in bone test involves insertion of an implant into the bone tissue of test animals. The biological response of the implant test sample is then compared with the biological response to implants control samples.

Test samples

The preparation of test and control samples are described in ISO 10993-6. Implant sizes are based on the size of the test animal and bone chosen. The following typical dimensions shall be considered for implants in mid-shaft cortical bone:

- Rabbits: cylindrical implants of 2 mm in diameter and 6 mm in length;
- Dogs, sheep, and goats: cylindrical implants of 4 mm in diameter and 12 mm in length; and
- Rabbits, dogs, sheep, goats, and pigs: 2 mm to 4.5 mm orthopedic bone screw-type implants;

Test animals and implant sites

Implants should be inserted into the bone of rodents, dogs, sheep, goats, pigs, or rabbits. The appropriate animal species should be selected as per provisions laid out in ISO 10993-2. Differences in bone physiology between different species should be assessed before the implantation procedures are initiated. It is also important to note that bone quality and bone densitometry may differ between species. For this reason, selection of suitable test animals should be justified and documented. Equivalent anatomical sites should be used for both the test and control samples. At least 10 test samples and 10 control samples should be evaluated for each implantation period.
A non-absorbable control sample shall be evaluated at each time point. A single time control point is acceptable provided an acceptable scientific justification is documented, which shall address the following:

Implantation procedure and evaluation

Bone preparation should be performed using low drilling speed and/or intermittent drilling with profuse irrigation with physiological saline solution and suction. The cortex of each femur or tibia should be exposed and the appropriate number of holes should be drilled to receive implants. For rabbits, up to three holes should be prepared while up to six holes should be prepared for larger animals. Implants should then be appropriately inserted. It is important that the diameter of the implant and the implant bed in the bone match well enough to avoid ingrowth of fibrous tissue. The implantation period should be determined as per ISO 10993-6 and evaluation of responses shall be observed and documented as per ISO 10993-6.

Box 18: Overview of Implantation Testing in Brain Tissue

Implantation in brain tissue

This test method is ideally used for assessing the biological response of brain tissue to an implanted test material. The implant is inserted into the neural tissue of test animals and the biological response to implant test samples is compared with the biological response of implant control samples.

Test samples

The preparation of test and control samples are described in ISO 10993-6. Implant sizes should be based on the animal species and site chosen. The following typical dimensions should be considered for rats and rabbits:

- For rod or wedge-shaped implants: a 1 mm × 1 mm or less in diameter/ section and 2 mm to 6 mm in length;
- An 8 mm diameter disc.

Test animals and implant sites

This test method involves the use of rats or rabbits. If other species are used, modifications of the testing protocol may be necessary. Both sexes should be represented in equal numbers and animals that are healthy and which have not been subjected to previous experimental procedures should be used. Species and age are important factors that should be taken into consideration. There should be a minimum of 8 tests and 8 negative control sites at each time point. These should be split equally between male and female animal species. Equivalent anatomical sites shall be used for the test and control samples ad careful selection of the implant site and surgical procedure is needed to minimize the risk of mechanical trauma. Additional guidance regarding implant sites is provided in ISO 10993-6.

Implantation procedure and evaluation

Each test animal should be weighed prior to implantation and again at periodic intervals thereafter. Species should undergo appropriate analgesia and anesthesia before the skull is prepared. Animals should be appropriately restrained during the operation procedure. Aseptic surgical techniques should be used. The skull should be exposed and holes of sufficient diameter should be made in the cranium so as to allow the implant samples in. In addition, a small hole is prepared in the dura and the implant is gently introduced into the appropriate part of the brain.

It is important to remember that the surgical technique employed will influence the result of any brain implantation procedure. A one-week implantation period is necessary, as well as other appropriate time intervals, to adequately characterize the response to the implant. Post implantation observation should be carried out and both microscopic and macroscopic changes should be observed and recorded.

Further guidelines are provided in ISO 10993-6.

3.10. HEMOCOMPATIBILITY

Blood forms one of the most complex biochemical systems within living organisms. It is essential for life functions such as the transport of oxygen, destruction of invading pathogens and the repair of damaged tissue. All of these form critical functions and for this reason, those medical devices that come into contact with blood must be hemocompatible and not adversely interact with any blood components. The ISO 10993-4 standard (Selection of tests for interactions with blood) describes a range of test categories for assessing interactions of medical devices that have either direct or indirect contact with blood.

Tests are recommended based on the type of device as defined in Table 3.30.

Table 3.30. Circulating Blood Contact Devices or Device Components and the Categories of Appropriate Testing for Consideration-External Communicating Devices and Implant Devices During Hemocompatibility Testing

Device examples	**Test Category**						
	Hemolysis		**Thrombosis**				
	Material-Induced	**Mechanically Induced**	**Coagulation**	**Platelet Activation**	**Complemented**	**Hematology**	***In vivo/ ex vivo*[a]**
External communicating devices							
Blood monitors (temporary/*ex vivo*)[b]	X		X	X		X	
Blood storage and administration equipment (e.g., infusion/transfusion sets), blood collection devices, extension sets	X		X	X		X	

Catheters in place for less than 24 h (e.g., atherectomy devices, intravascular ultrasound catheters, antegrade/retrograde coronary perfusion catheters, guide wires); cannulae	X		X[c]	X[c]		X[c]	X[c]
Catheters in place for more than 24 h (e.g., parenteral nutrition catheters, central venous catheters); cannulae	X		X[c]	X[c]		X[c]	X[c]
Cell savers[b]	X		X	X			
Devices for adsorption of specific substances from blood[b]	X	X	X	X	X		
Donor and therapeutic aphaeresis equipment and cell separation systems[b]	X	X	X	X	X		
Cardiopulmonary bypass system[b]	X	X	X[c]	X[c]	X	X[c]	X[c]
Hemodialysis/ hemofiltration equipment[b]	X	X	X[c]	X[c]	X	X[c]	X[c]
Leukocyte removal filter[b]	X	X	X[c]	X[c]	X	X[c]	X[c]
Percutaneous circulatory support devices[b]	X		X[c]	X[c]	X	X[c]	X[c]
Implant devices							
Annuloplasty rings, mechanical heart valves	X	X					X
Embolization devices	X						X
Endovascular grafts	X						X

Implantable defibrillator and cardioverter leads	X						X
Intra-aortic balloon pumps [b]	X	X					X
Pacemaker leads	X						X
Prosthetic (synthetic) vascular grafts and patches, including arterio-venous shunts	X						X
Stents (vascular)	X						X
Tissue heart valves, vascular grafts and patches and AV shunts	X						X
Total artificial hearts	X	X					X
Vena cava filters	X						X
Ventricular-assist devices [b]	X	X					X

a: Thrombosis is an in vivo or ex vivo phenomenon, but can be simulated with in vitro conditions. In vivo or ex vivo testing might not be necessary if clinically relevant in vitro thrombosis testing is performed.

b: Direct or indirect blood-contacting components only. For components that have only indirect blood contact, in vivo thrombogenesis and mechanical hemolysis or complement activation might not be necessary.

c: It is recognized that coagulation, platelet, and leucocyte responses are primarily involved in the process of thrombosis. Therefore, it is up to the manufacturer to decide if specific testing in the coagulation, platelet, and hematology test categories is appropriate as an alternate to in vivo testing.

d: See also ISO/TS 10993-20 for information on when complement activation should be considered for other end points such as anaphylaxis.

These tests can be divided into different categories depending on the primary process or system being measured. These include hemolysis and thrombosis (Table 3.31).

Table 3.31. Common Tests Used to Assess Interactions with Blood

Test by Categories		
Hemolysis	**Material Induced**	
	Mechanically induced	
Thrombosis	*In vitro*	Coagulation (Thrombin, fibrin assays, PTT assay)
		Platelet activation (Platelet count, indicator of activation (release products or platelets surface markers such as βTG, PF4, TxB2), platelet morphology
		Complement testing
		Hematology (Complete blood count (CBC), leucocyte activation)
	In vivo/ex vivo (Gross analysis, percentage occlusion, light microscopy, SEM)	–

Testing can be broken down into three main categories, namely *in vitro*, *in vivo,* and *ex vivo*. *In vitro* testing should ideally consider those designs that will simulate the anticipated worst-case clinical use condition. Tests should be carried out usually within 4 hours of drawing blood as some properties change quite rapidly after collection. Common variables that should be considered when using *in vitro* test methods include anticoagulant (type and amount), test sample preparation, test sample age, blood/blood component age, test sample storage, aeration, and pH, temperature, proper randomization, test sample surface area to blood volume ratio and for dynamic studies, fluid flow conditions, especially flow rate, wall shear rate and pressure(s). *Ex vivo* testing is generally carried out when the intended use of the device is, as the name suggests, *ex vivo*. A common example would be any external communicating device. *Ex vivo* test systems are used for monitoring various factors such as platelet adhesion, emboli generation, fibrinogen deposition, thrombus mass, white-cell adhesion, platelet consumption and platelet activation to name a few. *In vivo* testing on the other hand involves implanting the material or device in animals. Such *in vivo* test models are expected to be quite diverse, in order to appropriately mimic each clinical application. Common examples of such devices that can be tested by *in vivo* means include vascular patches, vascular catheters, vascular grafts, vascular stents, annuloplasty rings, heart valves and circulatory assist devices.

Animal models represent the closest simulation to actual use of the device prior to testing in humans. They allow for continuous device monitoring and a systematic controlled investigation of those important variables that may not necessarily be possible in *in vitro* and *ex vivo* testing. That being said, the choice of an animal model may be restricted by size requirements, the availability of certain animal species as well as cost. These factors, as well as others present limitations for testing and for this reason it is critical that the investigators be mindful of various differences between animal species. These include factors relating to coagulation, platelet functions and fibrinolysis, and the response to pharmacological agents such as anesthetics, anticoagulants, and so on. On the other hand, *in vitro* models also offer a great deal of benefits to testing hemocompatibility of medical devices because they avoid the high costs associated with animal models and the possibility of worst-case scenario testing.

CHAPTER 4

Understanding Your Medical Device

CONTENTS

4.1. SAMPLE PREPARATION: DETERMINING THE SURFACE AREA OF YOUR DEVICE

The standard surface area of the medical device is used to determine the volume of extract needed for each biocompatibility test performed. This area includes the combined area of both sides of the device but excludes unspecified surface irregularities. If, however, the surface area cannot be determined due to the shape or configuration of the device, a mass/volume of extracting fluid can be used.

The easiest method for determining the surface area of a device is to breakdown the device into basic geometric shapes. Often this is done by design engineers using appropriate software. A manufacturer may be asked to calculate the surface area of their device before sending the respective device off for testing. General equations to calculate surface area are presented in Table 4.1.

Table 4.1. Formulas for Surface Area Calculation

DEVICE SHAPE	Device	Formula
Square or rectangle	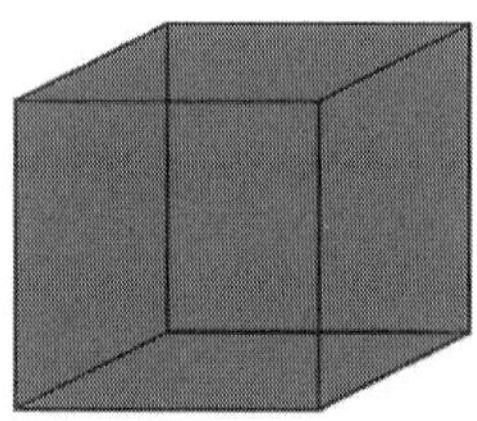	$A = L \times W$
Hollow cylinder	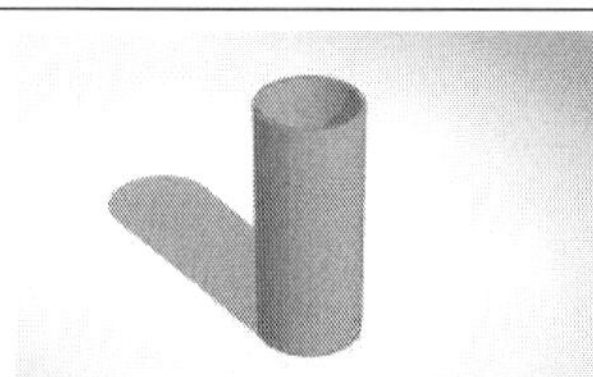	$A = (ID + OD)\, \pi \times L$
Disk		A (one side) $= \pi\, r^2$

Ellipse		A = (π × X × Y)/4
Regular polygon	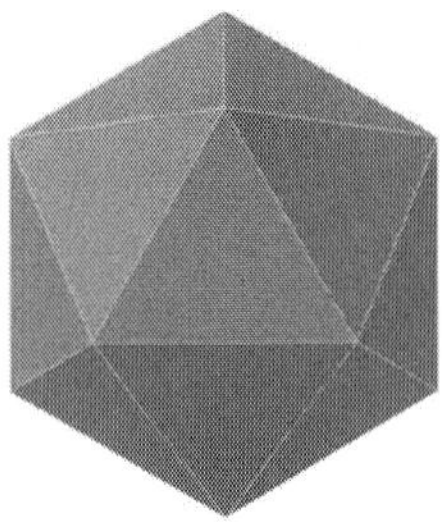	A = (b × h × n)/2
Solid cylinder (including ends)	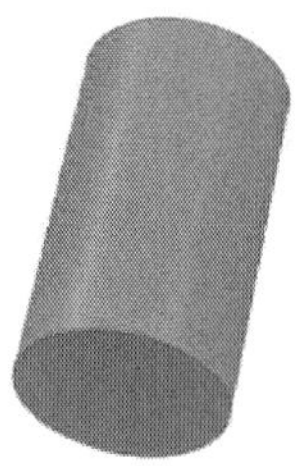	A = (OD × π × L) + (2 π r^2)
Triangle	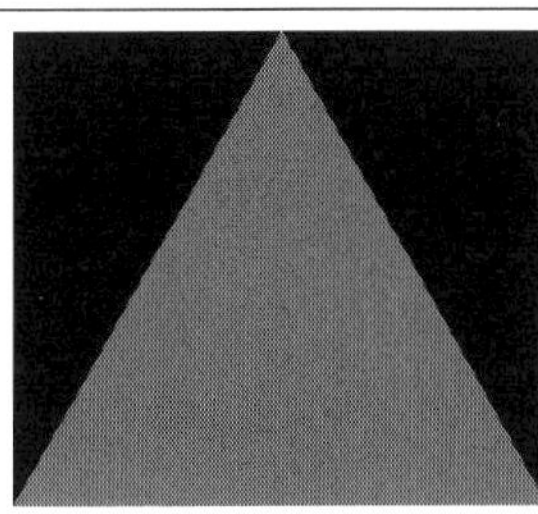	A = (b × h)/2
Sphere		A = 4 × π r^2

Trapezoid	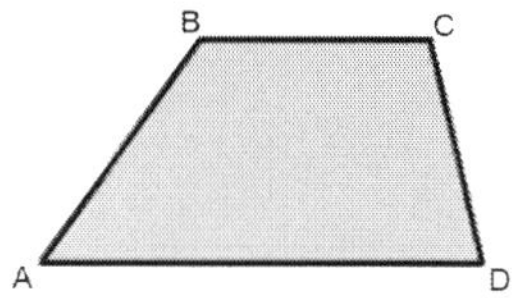	$A = (h \times [p + q])/2$
Circular ring	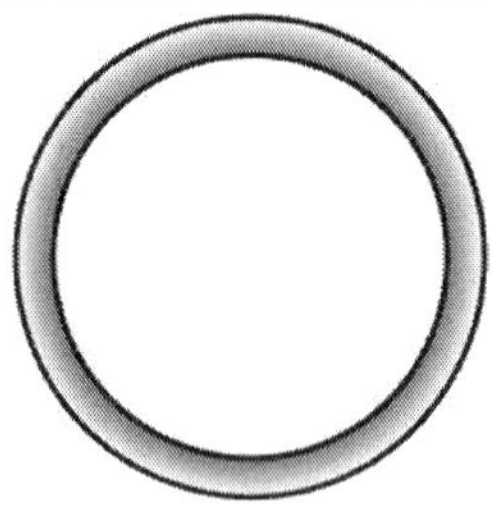	$4\ \pi^2 R_r r_c$

Legend:

A	Surface area
OD	Outer diameter
W	Width
R_R	Ring radius
X, Y	Longest and shortest distances through the center of an ellipse
h	Height
p, q	Length of the parallel side of a trapezoid
ID	Inner diameter
L	Length
R	Radius
r_c	Cross-section radius
π	3.14
b	Base length
n	Number of sides of a polygon
r_i	½ ID

The standard surface area can be used to determine the volume of extraction volume needed. This includes the combined area of both sides of the sample to be tested and excludes unknown or unspecified surface irregularities. The device should be cut into small pieces before extraction to enhance exposure and submersions to the extracting media where appropriate. Common examples of medical devices that would be cut up during biocompatibility testing include those made of polymers. Devices constructed of elastomers, coated materials, composites or laminates shall be tested intact and whole to avoid differences in extraction characteristics that may arise when such surfaces are cut.

If a device is made of multiple materials and cannot be tested as a whole, then each individual material in the final product shall be represented proportionally in the test sample. For example, if a device has a surface coating, then both the surface coating and device substrate shall be included for testing, even if the substrate has no tissue contact. In addition, the potential for interactions of different materials or synergistic effects should also be considered.

4.2. CHOOSING THE APPROPRIATE EXTRACTION MEDIA AND EXTRACTION CONDITIONS

Most medical devices are made of varying materials such as plastics, metals, and various other solid materials. Biocompatibility problems associated with medical devices are often caused by toxins that leach out of the device itself and into surrounding tissues and/or body fluids. For most tests, the preparation of fluid extracts from the device materials tend to be the most appropriate technique to provide test samples for determining the biological reactivity of any toxins that may leach out.

In much of the recommended biocompatibility tests specified by ISO 10993, extracts from the materials of the device are used in assessing biocompatibility. According to the ISO 10993-12 standard for Sample Preparation and Reference Materials, the extraction conditions should be appropriate to the final device and its end use. This means that it is critical that the extraction medical selected or testing represent the environment in which the final product will be used. Extraction should be done using both polar and non-polar vehicles. Commonly used extraction media include sodium chloride, vegetable oil, alcohol, polyethylene glycol (PEG), dimethyl sulfoxide (DMSO), and clinically relevant solvents (Table 4.2).

Table 4.2. Commonly Used Extraction Vehicles

Extraction Vehicle	Examples
Polar extraction vehicles	Water, physiological saline, culture media without serum

Non-polar extraction vehicles	Refined vegetable oil such as cottonseed or sesame oil
Other extraction vehicles	Ethanol/water, ethanol/saline, polyethylene glycol 400, dimethyl-sulfoxide, culture media with serum

The appropriate extract is often selected on the basis of the biological environment in which the test material is to be used. For example:

- A saline extract mimics the aqueous, hydrophilic fluids in the body and allows for the use of extreme temperatures in preparing extracts, thus simulating certain sterilization conditions.
- Tissue culture media may even more closely mimic aqueous body fluids, but on the other hand, cannot be used for high temperature extractions.
- Vegetable oils are non-polar, hydrophobic solvents and simulate the lipid fluids in the body.
- DMSO extracts are often used in certain genotoxicity and sensitization tests.

Alcohol and PEG should be used only if they approximate the solvent properties of drugs or other materials that will come into contact with the device during its normal use.

4.2.1. Extraction Conditions and Methods

Extraction conditions are generally defined by temperature and time as per the ISO 10993 standard and justified on the basis of providing a standardized approach. These conditions should be at a minimum, as extreme as any conditions the device or material being tested will encounter. This can be either during sterilization or clinical use. In terms of temperature, the highest extraction temperature that does not melt or fuse the material or cause chemical changes should be chosen. Common extraction conditions are listed below:

- 37°C for 72 hours
- 50°C for 72 hours
- 70°C for 24 hours
- 121°C for 1 hour

For medical devices which are in short term contact with intact skin or mucosa and which are not implanted, extraction times of less than 24 hours (but not less than 4 hours) are acceptable. For most test materials, extractions are performed under static conditions; however, agitation may be used if deemed necessary to more closely mimic the end use of the device or to alternatively ensure that the extraction media comes into contact with the relevant device components being tested.

The standard surface area can be used to determine the volume of extraction vehicle needed. The amount of test material used during the extraction process is usually determined as a ratio of sample surface area to extractant volume or sample mass to extractant volume generally, the surface area ratio should be used wherever possible, with a mass-to-volume ratio being used only for testing of irregularly shaped devices. Extraction ratios specified within the ISO 10993-12 standard are outlined in Table 4.3.

Table 4.3. Extraction Ratios for Medical Devices as per ISO 10993-12: Biological Evaluation of Medical Devices (Sample preparation and reference materials)

Thickness (mm)	**Extraction Ratio (Surface Area or Mass/Volume)** ± 10%	**Examples of Forms of Materials**
< 0.5	6 cm^2/ml	Film, sheet, tubing wall
0.5 to 1.0	3 cm^2/ml	Tubing wall, slab, small molded items
< 1.0	3 cm^2/ml	Larger molded items
< 1.0	1.25 cm^2/ml	Elastomeric closures
Irregularly shaped solid devices	0.2 cm^2/ml	Powder, pellets, foam, non-absorbent molded items
Irregularly shaped porous devices (low density materials)	0.1 cm^2/ml	Membranes, textiles

The standard also identifies that there are no standardized methods available for testing absorbents and hydrocolloids. For this reason, it suggests the following protocol:

- Determine the volume of extraction vehicle that each 0.1 g or 1.0 cm^2 of material absorbs;
- Then, in performing the material extraction, add this additional volume to each 0.1 g or 1.0 cm^2 in an extraction mixture.

Box 19: Important Principles to Remember for test Sample Extraction

- Where extracts of a medical devices are prepared, the extraction vehicle and conditions of extraction used should be suitable to the nature and use of the final product.
- It should be remembered that the purpose of such biological tests is to identify hazards and estimate the risks of the hazards occurring in both exaggerated use and/or in actual conditions of use.
- the amount of extractable substance(s) is related to the period of extraction, temperature, the ratio of material surface area to volume of extraction vehicle and the overall nature of the extraction vehicle.
- The period of extraction should be sufficient to maximize the amount of material extracted.
- Extraction temperatures may vary for the different materials to be tested.
- Extraction should not cause significant degradation of the material, unless the material is intended to dissolve or be resorbed during use.
- The extraction temperature is dependent upon the physicochemical characteristics of the device material(s).
- The ratio of the surface area of the device to the volume of extraction vehicle or solvent should be sufficient to:

i. attain the maximum amount of extractable substance(s) in an appropriate dosage volume for biological testing or chemical analysis
ii. demonstrate the hazard potential of using the device in humans
ii. cover the material in the solvent volume.

- The solvent(s) selected as the extraction vehicle should:

i. Be suitable for use in the specific biological test systems
ii. Simulate the extraction which occurs during clinical use of the device
iii. Maximize the amount of extractives

- The use of standard polar and non-polar solvents is recommended
- Follow the conditions prescribed as per ISO 10993-12 for materials that dissolve or resorb in the body

4.2.2. Choosing the Appropriate Containers for Extraction

The extraction should be performed in a clean and chemically inert container. This should be closed and have minimal dead space. Usually an extraction container made of borosilicate glass or any other form of inert material should be used. ISO 10993-12 provides further guidelines on the types of containers that can be used for extraction.

4.3. SELECTING SUITABLE REFERENCE MATERIALS AND EXPERIMENTAL CONTROLS

Reference materials (RM) or certified reference materials (CRM) shall be used in biological tests as controls. The purpose of this to show the suitability of a procedure to yield and known and reproducible response. RM used as controls should be in the same material class as the test sample being tested. These RM should also be handled in the sample way as test samples are handled and prepared for the testing. This includes suitable handling to prevent contamination and cutting into pieces where needed.

CRM are generally selected because of their high purity and various other critical chemical, physical, and biological characteristics. These characteristics are usually determined and confirmed by collaborative testing in three or more laboratories. The data is usually made available to the investigator. Suppliers of such materials should certify these materials and ideally provide a reference certificate confirming such characteristics.

Table 4.4. Suitable Positive and Negative Controls

Type of Control	Examples
Negative controls	• high-density polyethylene • low-density polyethylene • silica-free polydimethylsiloxane • polyvinylchloride • polyether urethane • polypropylene • aluminum oxide ceramic rods • stainless steel • commercially pure titanium alloys
Positive controls	• polyvinylchloride-containing organotin • additives • segmented polyurethane rod or films containing zinc diethyldithiocarbamate (SPU-ZDEC) or dibutyl dithiocarbamate (SPU-ZDBC) • certain latex formulations • solutions of zinc salts • solutions of copper salts • dilutions of phenol and water

CHAPTER 5

How to Set Up a Biocompatibility Testing Laboratory

CONTENTS

5.1. REQUIREMENTS OF A LABORATORY

A biocompatibility testing laboratory can be broken down into eight major areas of testing. When setting up a lab, depending on the area of specialty, some may choose to set up individual laboratories specialized in particular form of testing or a general laboratory which covers various types of biocompatibility testing. For the purpose of this chapter, we will break down the lab set up per biocompatibility testing type. The following laboratory set-ups will be focused on:

- Cytotoxicity testing laboratory (Tissue Culture Laboratory);
- Tissue compatibility evaluation laboratory;
- Hemocompatibility evaluation laboratory;
- Sterility evaluation laboratory;
- Histopathology evaluation laboratory;
- Accelerated Aging laboratory;
- Physiochemical evaluation laboratory; and
- Package validation laboratory.

5.2. CYTOTOXICITY TESTING LABORATORY

Cytocompatibility or cytotoxicity tests are considered as screening assays that are used to evaluate a living cells reaction to a medical device or components within a medical device. This can include a cells viability as well as the overall ability for cellular growth. This form of testing involves the exposure of substances extracted from a test material to a particular cell culture line. Cell cultures are known to be extremely sensitive to minute quantities of leachable chemicals. These cell lines then exhibit characteristic signs of toxicity in the presence of these potentially harmful leachables. This type of testing is commonly used to determine if a particular type of material is safe or as a periodic check for routinely used materials so as to ensure that no shift in quality has occurred. Usually, a cell suspension of L-929 mouse fibroblasts are used to seed petri dishes that contain a known concentration of cells supplemented with minimal essential medium (MEM). The cells are then incubated at 37°C for a period between 24–72 hours and observed daily for cell culture confluency. Cells are scored for reactivity on a grading scale as per the ISO 10993-5 standard. If the biomaterial extract shows no evidence of causing cell lysis or toxicity, and if the cell cultures exposed to the test item do not show greater than a mild reactivity, then the biomaterial is considered to be cytocompatible.

Testing can be broken down into qualitative and quantitative tests.

There are three qualitative cytocompatibility evaluation tests commonly used for medical devices:

1. The direct contact test;
2. The agar diffusion assay; and
3. The MEM elution assay.

1. **The Direct Contact Test:** The Direct Contact test is recommended for low-density materials, such as contact lens polymers. In this method, a piece of test material is placed directly onto cells growing on a culture medium. The cells are then incubated. During incubation, leachable chemicals in the test material can diffuse into the culture medium and contact the cell layer. Reactivity of the test sample is indicated by malformation, degeneration, and lysis of cells around the test material.
2. **The Agar Diffusion Assay:** The agar diffusion assay is usually recommended for high-density materials, such as elastomeric closures. This test method involves a thin layer of nutrient-supplemented agar which is placed over cultured cells. The test material (or an extract of the test material which has been dried on filter paper) is placed on top of the agar layer. The cells are then incubated. After incubation, a zone of malformed, degenerative, or lysed cells will have formed under and around the test material, indicating cytotoxicity.
3. **The MEM Elution Assay:** The MEM Elution assay uses different extracting media and extraction conditions to test devices according to actual use or exaggerated conditions. Extracts can then be titrated to yield a semi-quantitative measurement of cytotoxicity. After preparation, the extracts are then transferred onto a layer of cells and incubated. Following incubation, the cells are examined microscopically formal formation, degeneration, and lysis of the cells.

There are two quantitative tests commonly used, namely:

a. The MTT assay; and
b. The colony formation assay.

   a. **The MTT Assay:** The MTT assay measures the viability of cells by spectrophotometric analyses. This test measures the reduction of the yellow, water-soluble MTT (3-4,5-dimethyl-thiazol-2-yl) – (2,5-diphenyl tetrazolium bromide) by mitochondrial succinate dehydrogenase. This biochemical reaction is only catalyzed by living cells and a minimum of four concentrations of the test material are tested.

b. **The Colony Formation Assay:** The colony formation assay enumerates the number of colonies formed after exposing them to the test material. This exposure is done at different concentrations. It is a very sensitive test as colony formation is assessed while the cells are still in a state of proliferation. This phase is chosen as they are more susceptible to toxic effects at this point. Results can be analyzed by determining a concentration-dependence curve which evaluates the induced inhibition of the test material. In addition, the IC50 value (concentration of the test material that provides 50% inhibition) can be calculated.

Ideally, at least one type of cytotoxicity test-qualitative and quantitative, should be performed on each component of a device. An overview of the cytotoxicity tests and their relevant standards which should serve as a guide are given in Table 5.1.

Table 5.1. Overview of Cytotoxicity Evaluation Tests

Type of Testing	**Test Type**		**Test Standard**
Cytocompatibility/ cytotoxicity	*In vitro* cytotoxicity	Direct contact	ISO 10993-5
		Indirect contact	
		Test on extract	
		MTT assay	
	Cell adhesion		

The basic equipment required to set up a basic cytotoxicity testing laboratory are highlighted in Table 5.2.

Table 5.2. Basic Equipment Required to Set Up a Cytotoxicity Testing Laboratory

Equipment	Definition	Example
Laminar flow	A laminar flow cabinet or tissue culture hood is a carefully enclosed bench designed to prevent contamination. Air is drawn through a HEPA filter and blown towards the user. The cabinet is usually made of stainless steel with no gaps or joints where spores might collect.	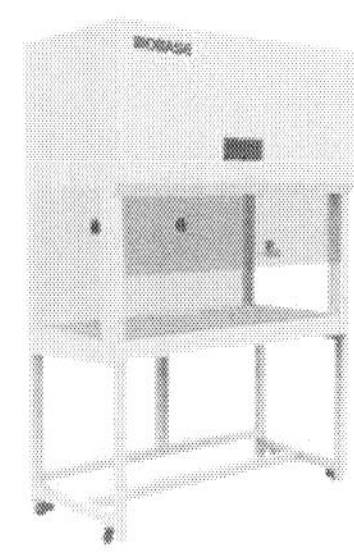
CO_2 incubator	An incubator is a device used to grow and maintain microbiological cultures or cell cultures. The incubator maintains optimal temperature, humidity, and other conditions such as the carbon dioxide (CO_2) and oxygen content of the atmosphere inside.	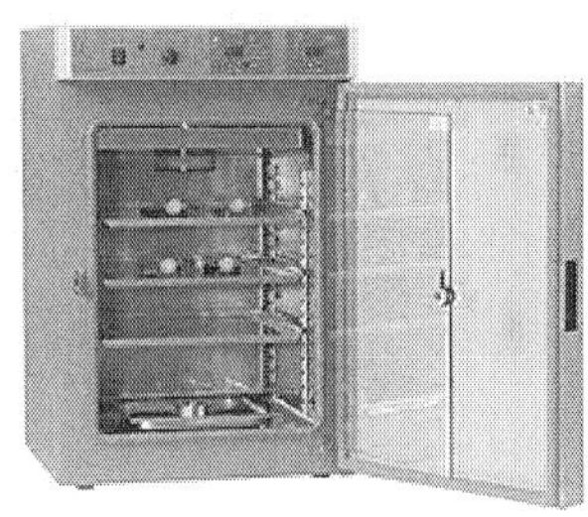
Binocular microscope	A microscope is an instrument used to see objects that are microscopic to the naked eye. A binocular microscope has two eyepieces.	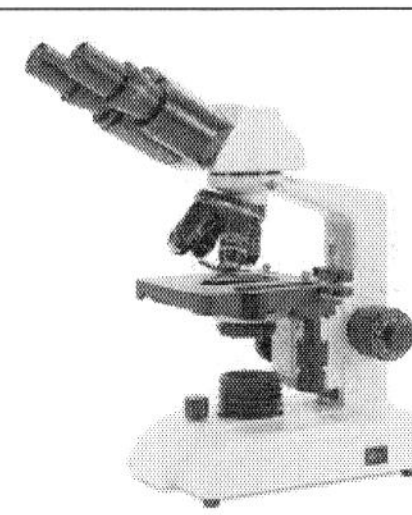
Autoclave	An autoclave is a device used to sterilize equipment and supplies by subjecting them to high pressure saturated steam at 121°C for around 15–20 minutes.	

Sterilizer (Ethylene oxide)	Ethylene Oxide (EtO) sterilization is mainly used to sterilize medical and pharmaceutical products that cannot support conventional high-temperature steam sterilization. This is commonly used for devices that incorporate electronic components, plastic packaging or plastic containers.	
Pipettes	A pipette is a laboratory tool commonly used to transport a measured volume of liquid.	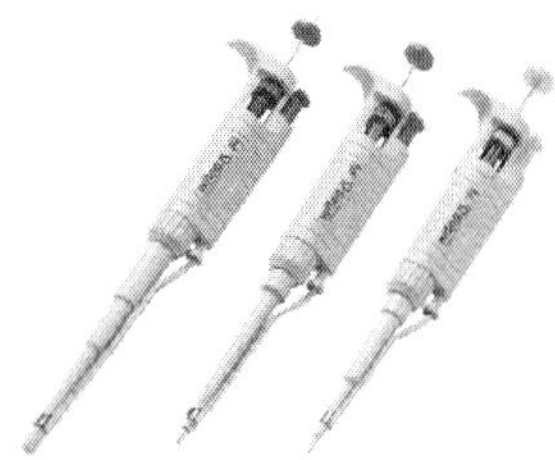
Laboratory water bath	A water bath is a piece of equipment used to heat materials gently and gradually to fixed temperatures, or to keep materials warm over a period of time.	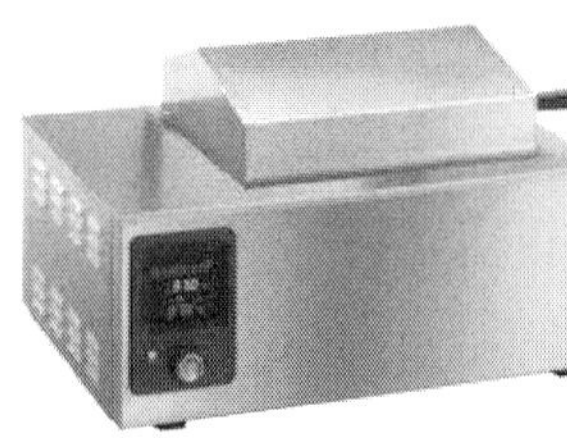
Cryogenic storage vessel	Cryogenic storage vessels are specialized types of vacuum flask used for storing cryogens (such as liquid nitrogen or liquid helium), whose boiling points are much lower than room temperature.	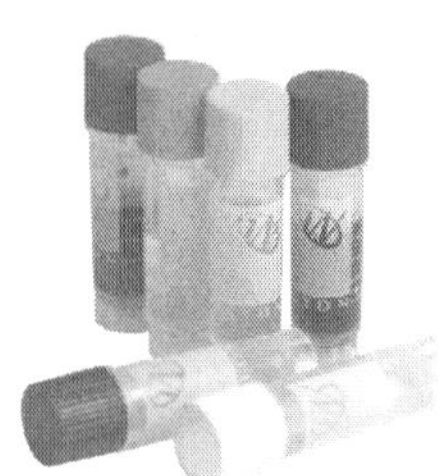
Basic laboratory glassware/plasticware	Laboratory Glassware forms a basic part of scientific laboratories. Ideal laboratory glassware must comprise of borosilicate glass with a low coefficient of expansion, for resistance to heat, and a very high resistance to chemical attack.	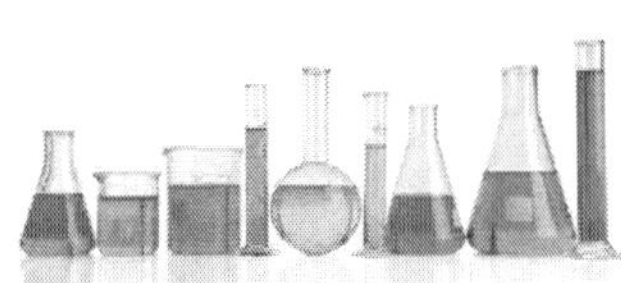

Ultraviolet lamp	Ultraviolet radiation is widely used for killing bacteria or producing fluorescence.	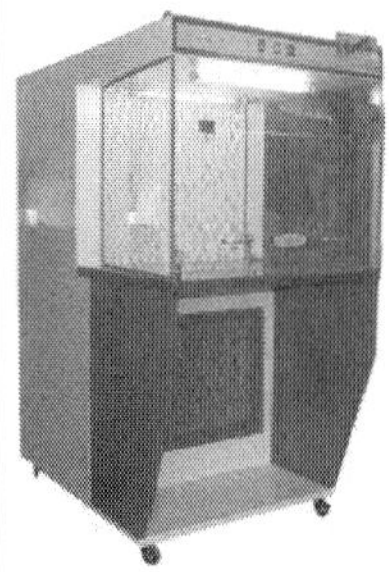
Air monitoring system	A suitable air monitoring system should be in place which allows for laboratory air temp and humidity to be monitored.	

5.3. TISSUE COMPATIBILITY EVALUATION LABORATORY

A tissue compatibility laboratory involves testing which estimates the local irritation potential of devices, materials or extracts. This is usually done using sites such as skin or mucous membranes in an animal model. The route of exposure (skin, eye, mucosa) used during testing as well as the duration of contact should be similar to the anticipated clinical use of the device. In certain instances, some manufacturers may even request a more exaggerated exposure to establish a margin of safety for patients.

The Intracutaneous Test

This test aims to detect potential local irritation due to chemicals that may get extracted from a biomaterial. It involves extracts of the test material and blanks which are injected intradermally. The injection sites are scored for any redness or swelling, also known as erythema and edema. This procedure is generally recommended for devices that will have internal contact with the body or body fluids.

The Primary Skin Irritation Test

The primary skin irritation test is generally considered for topical devices that have external contact with intact or breached skin.

The Mucous Membrane Irritation Test

This test is generally recommended for devices that will have contact with natural channels or tissues and often use extracts rather than the material itself. Common procedures include vaginal, cheek pouch and eye irritation studies.

An overview of the tissue compatibility tests and their relevant standards which should serve as a guide are given in Table 5.3.

Table 5.3. Overview of Tissue Compatibility Tests and Relevant Standards

Type of Testing	Test Type		Test Standard
Tissue compatibility	Irritation test	Intracutaneous irritation test (Rabbit) on cotton seed oil and normal saline extract	ISO 10993-10
		Skin Irritation test	
		Vaginal irritation test	
		Penile irritation test	
	Acute systemic toxicity	Intravenous/intraperitoneal on cotton seed oil and normal saline extract	ISO 10993-11
	Sensitization	Closed patch method-Guinea Pig	ISO 10993-10
		Maximization method-Guinea Pig	
	Rabbit Pyrogen Test		ISO 10993-11
	Hemolytic property (Rabbit Blood)		ASTM 756
	Implantation test		ISO 10993-6
	Genotoxicity test		ISO 10993-3
	Mucous membrane irritancy test in rabbit	Acute eye irritation test in rabbit/corrosion	

The basic equipment required to set up a basic tissue compatibility evaluation laboratory are highlighted in Table 5.4.

Table 5.4. Basic Equipment Required to Set Up a Basic Tissue Compatibility Evaluation Laboratory

Equipment	**Definition**	**Example**
Laminar flow	A laminar flow cabinet or tissue culture hood is a carefully enclosed bench designed to prevent contamination. Air is drawn through a HEPA filter and blown towards the user. The cabinet is usually made of stainless steel with no gaps or joints where spores might collect.	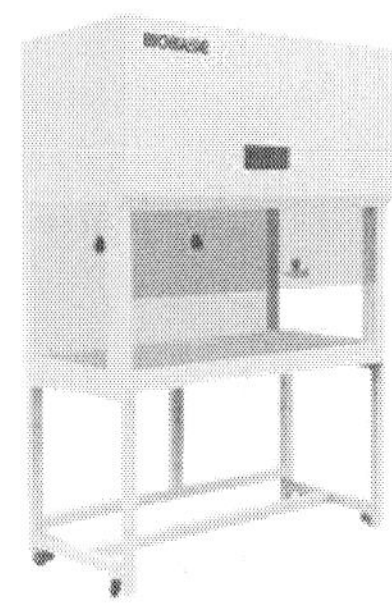
CO_2 incubator	An incubator is a device used to grow and maintain microbiological cultures or cell cultures. The incubator maintains optimal temperature, humidity, and other conditions such as the carbon dioxide (CO_2) and oxygen content of the atmosphere inside.	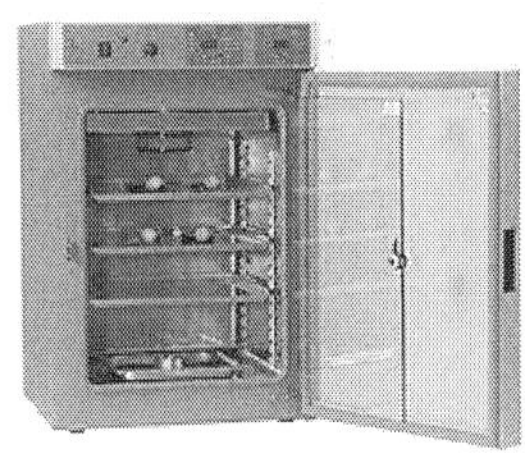
Binocular microscope	A microscope is an instrument used to see objects that are microscopic to the naked eye. A binocular microscope has two eye pieces.	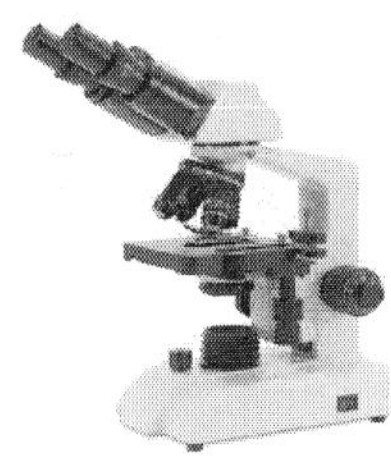
Autoclave	An autoclave is a device used to sterilize equipment and supplies by subjecting them to high pressure saturated steam at 121°C for around 15–20 minutes.	

Sterilizer (Ethylene oxide)	Ethylene Oxide (EtO) sterilization is mainly used to sterilize medical and pharmaceutical products that cannot support conventional high temperature steam sterilization. This is commonly used for devices that incorporate electronic components, plastic packaging or plastic containers.	
Pipettes	A pipette is a laboratory tool commonly used to transport a measured volume of liquid.	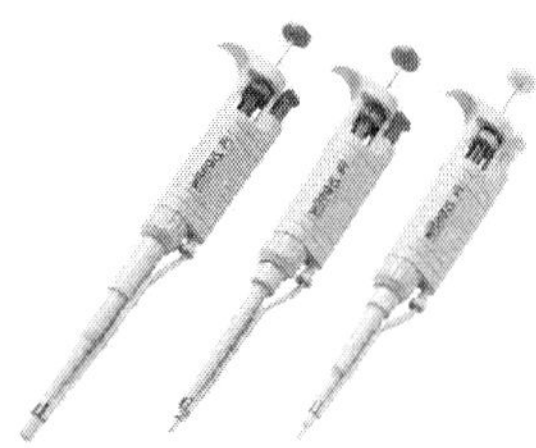
Laboratory water bath	A water bath is a piece of equipment used to heat materials gently and gradually to fixed temperatures, or to keep materials warm over a period of time.	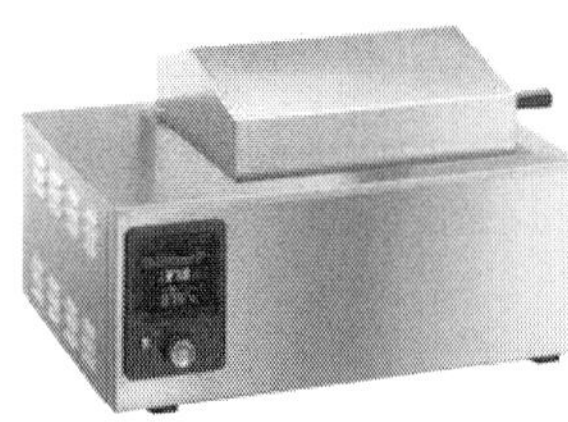
Cryogenic storage vessel	Cryogenic storage vessels are specialized types of vacuum flask used for storing cryogens (such as liquid nitrogen or liquid helium), whose boiling points are much lower than room temperature.	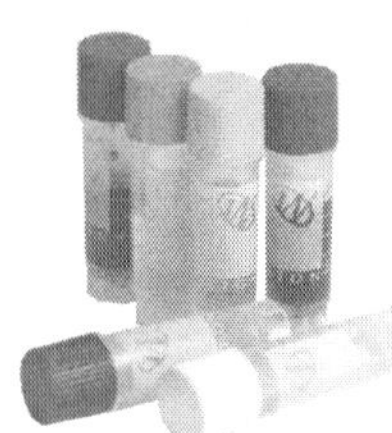
Basic laboratory glassware/plasticware	Laboratory Glassware forms a basic part of scientific laboratories. Ideal laboratory glassware must comprise of borosilicate glass with a low coefficient of expansion, for resistance to heat, and a very high resistance to chemical attack.	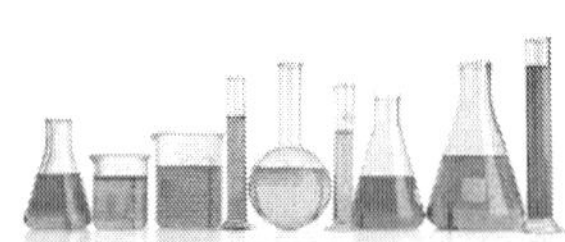

Ultraviolet lamp	Ultraviolet radiation is widely used for killing bacteria or producing fluorescence.	
Air monitoring system	A suitable air monitoring system should be in place which allows for laboratory air temp and humidity to be monitored.	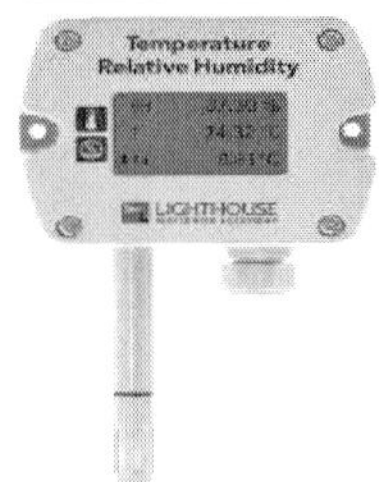
thermometer	A laboratory thermometer is used to measure temperatures with high accuracy. It can be partially or fully immersed in substance being measured.	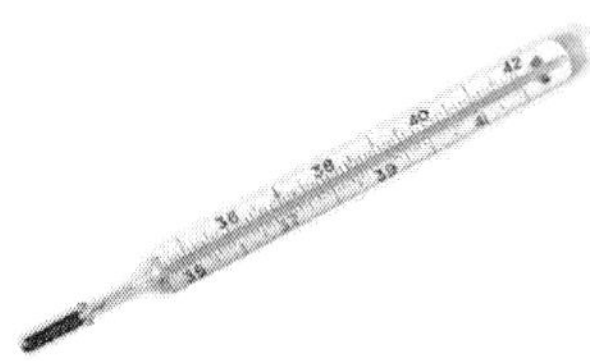
General laboratory instruments	This encompasses the various tools and equipment used by scientists on a day to day basis. This includes Bunsen burners, burettes, petri dishes, graduated cylinders, laboratory scales and so on	
Chemical testing instruments	Chemical testing equipment includes those pieces of equipment used to determine, characterize, and quantify chemical components. This can include spectroscopy systems, gas chromatography and liquid chromatography systems, process analyzers, thermal analyzers, calorimeters, and so on	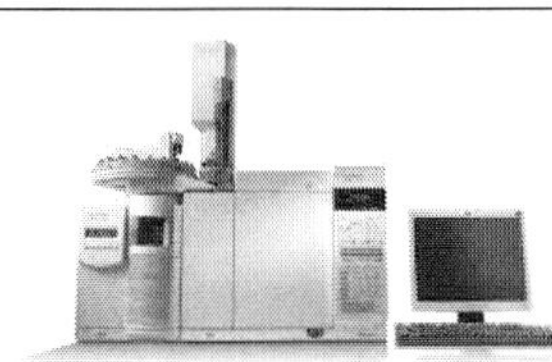

5.4. HEMOCOMPATIBILITY EVALUATION LABORATORY

A hemocompatibility evaluation laboratory focuses on the evaluation of material-to-blood interactions. The tests used for evaluation of blood compatibility are broken down into five categories based on the process or system being tested. This includes such as thrombosis, coagulation, platelets, and platelet function, hematology, and immunology. In addition, other indicators of blood compatibility are also evaluated. This includes quantitative estimation of protein absorption, endothelial cell adhesion as well as proliferation on surfaces with the ability to resist shear stress. These specialty labs may also focus on the development of new reagents and tests used for evaluation of blood-material interactions, the development of hemostatic bioadhesives as well as carry out studies on the modification of blood-contacting biomaterials so as to allow for reduction in thrombogenicity.

Table 5.5. Type of Hemocompatibility Tests That Can Be Carried Out in a Hemocompatibility Evaluation Laboratory

Type of Testing	Test Type	Test Standard
Hemocompatibility evaluation laboratory	Percentage hemolysis	ISO 10993-4
	Partial thromboplastin time (PTT)	
	Prothrombin time (PT)	
	Fibrinogen assay	

The basic equipment required to set up a basic hemocompatibility evaluation laboratory are highlighted in Table 5.6.

Table 5.6. Basic Equipment Required to set up a Basic Hemocompatibility Evaluation Laboratory

Equipment	Definition	Example
Laminar flow	A laminar flow cabinet or tissue culture hood is a carefully enclosed bench designed to prevent contamination. Air is drawn through a HEPA filter and blown towards the user. The cabinet is usually made of stainless steel with no gaps or joints where spores might collect.	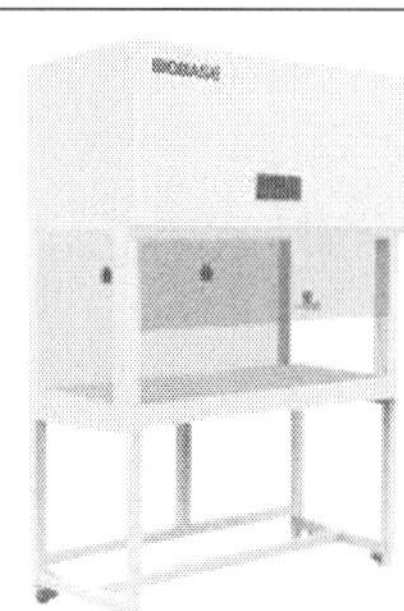

CO_2 incubator	An incubator is a device used to grow and maintain microbiological cultures or cell cultures. The incubator maintains optimal temperature, humidity, and other conditions such as the carbon dioxide (CO_2) and oxygen content of the atmosphere inside.	
Incubator	An incubator is a heated insulated box that maintains optimal temperature, humidity, gaseous content of the atmosphere inside. It is used to grow and maintain microbiological/cell cultures.	
Spectrophotometer (UV based)	An instrument for measuring the intensity of wavelengths in a spectrum of light, transmitted by a substance or a solution.	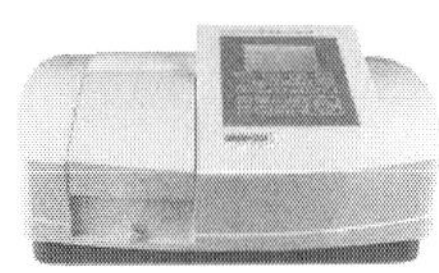
Binocular microscope	A microscope is an instrument used to see objects that are microscopic to the naked eye. A binocular microscope has two eye pieces.	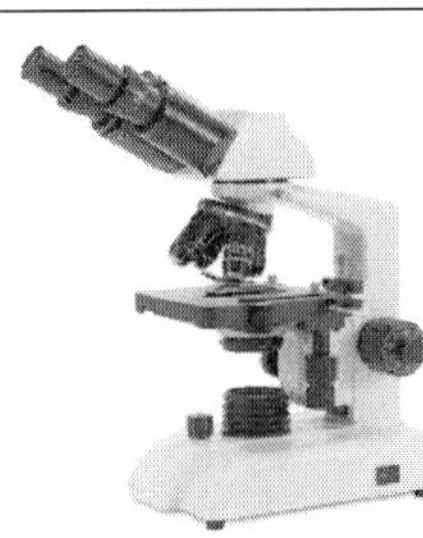

Autoclave	An autoclave is a device used to sterilize equipment and supplies by subjecting them to high pressure saturated steam at 121°C for around 15–20 minutes.	
Sterilizer (Ethylene oxide)	Ethylene Oxide (EtO) sterilization is mainly used to sterilize medical and pharmaceutical products that cannot support conventional high temperature steam sterilization. This is commonly used for devices that incorporate electronic components, plastic packaging or plastic containers.	
Pipettes	A pipette is a laboratory tool commonly used to transport a measured volume of liquid.	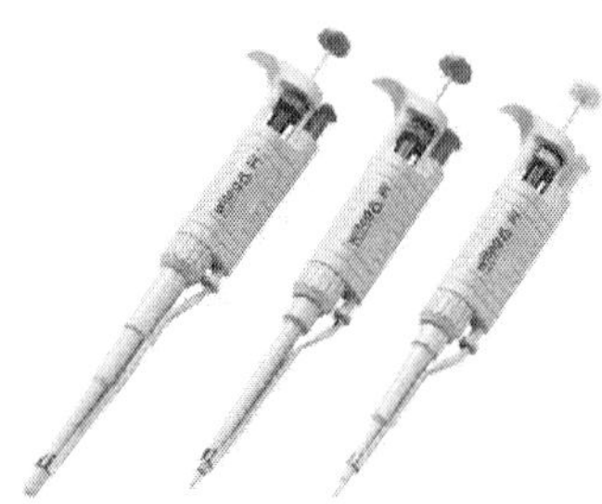
Laboratory water bath	A water bath is a piece of equipment used to heat materials gently and gradually to fixed temperatures, or to keep materials warm over a period of time.	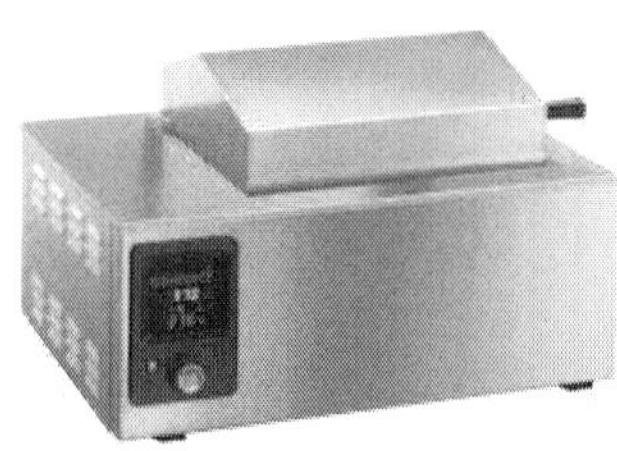

Cryogenic storage vessel	Cryogenic storage vessels are specialized types of vacuum flask used for storing cryogens (such as liquid nitrogen or liquid helium), whose boiling points are much lower than room temperature.	
Basic laboratory glassware/plasticware	Laboratory Glassware forms a basic part of scientific laboratories. Ideal laboratory glassware must comprise of borosilicate glass with a low coefficient of expansion, for resistance to heat, and a very high resistance to chemical attack.	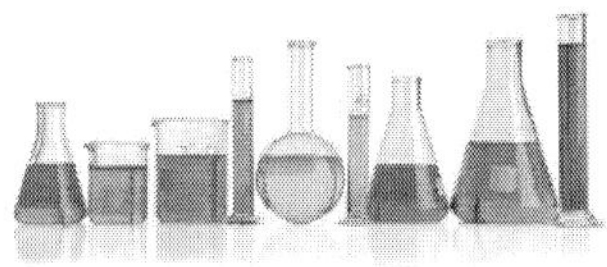
Ultraviolet lamp	Ultraviolet radiation is widely used for killing bacteria or producing fluorescence.	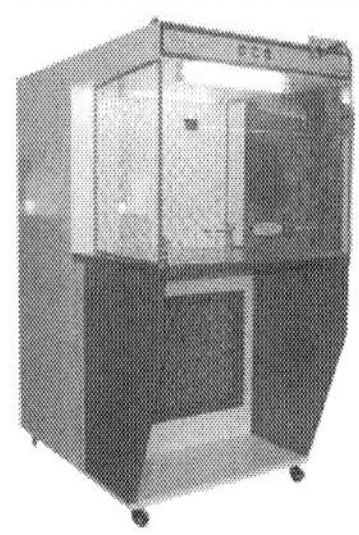
Air monitoring system	A suitable air monitoring system should be in place which allows for laboratory air temp and humidity to be monitored.	
thermometer	A laboratory thermometer is used to measure temperatures with high accuracy. It can be partially or fully immersed in substance being measured.	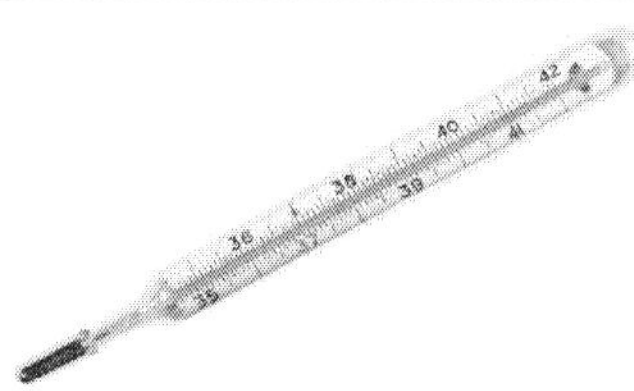

General laboratory instruments	This encompasses the various tools and equipment used by scientists on a day to day basis. This includes Bunsen burners, burettes, petri dishes, graduated cylinders, laboratory scales and so on	
Chemical testing instruments	Chemical testing equipment includes those pieces of equipment used to determine, characterize, and quantify chemical components. This can include spectroscopy systems, gas chromatography and liquid chromatography systems, process analyzers, thermal analyzers, calorimeters, and so on	
Biochemistry analyzer	This is an instrument that uses the serum of centrifuged blood or urine to biochemically measure various components, such as sugar, cholesterol, protein, enzymes, and so on.	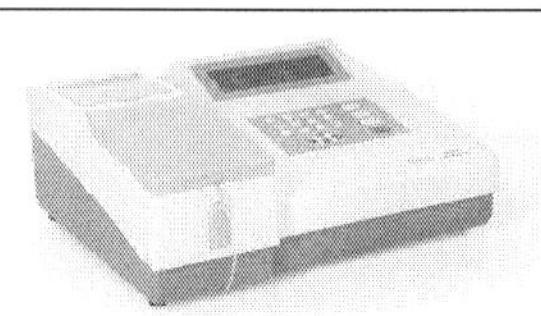
Hematology analyzer	This piece of equipment is used to conduct a complete blood count (CBC). This includes red blood cells (RBC), white blood cells (WBC), hemoglobin, and platelet counts.	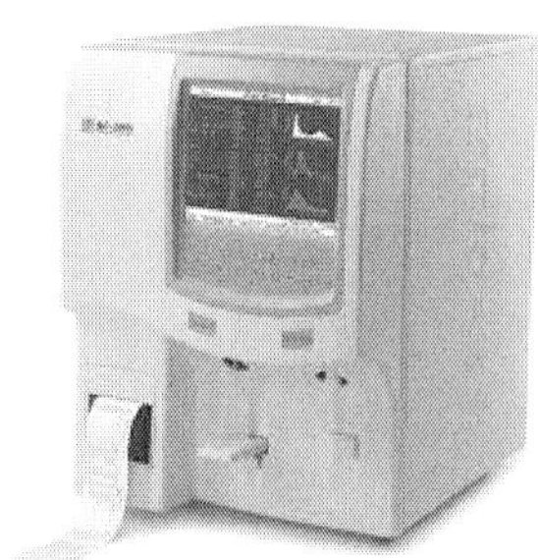
Coagulometer	An instrument used to detect the coagulation efficiency of blood	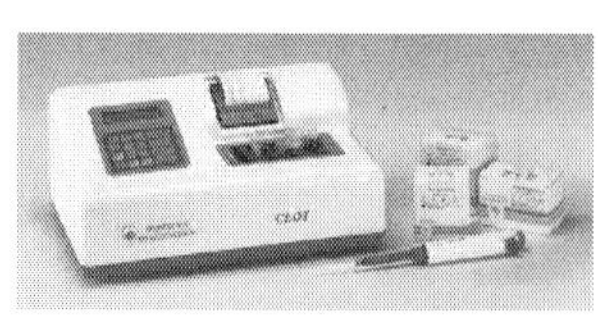

Aggregometer	An instrument used for measuring the aggregation of platelets in a sample of blood	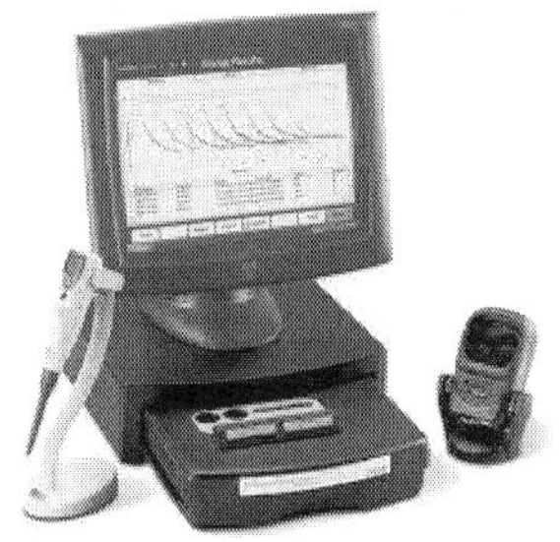
Elisa reader	Also known as micro titer plates, these instruments detect and process biological and chemical data using absorbance (ELISAs, enzyme activity, and nucleic acid and protein quantification), luminescence, and fluorescence detection modes, including intensity, TRF, and polarization.	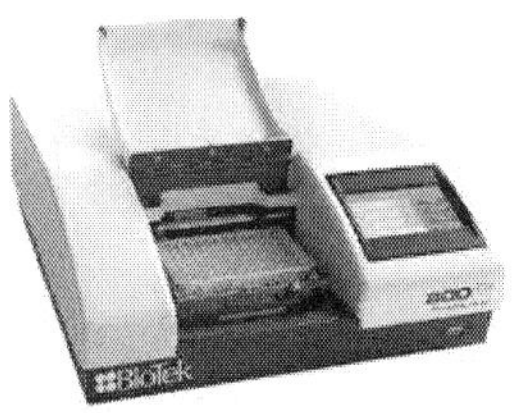
Flow cytometer	An instrument that uses laser-based technology to count and sort cells and detect biomarkers among other things. It is able to detect and measure physical and chemical characteristics of a population of cells or particles.	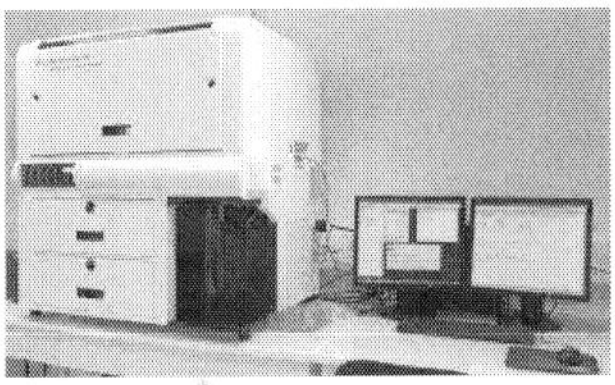
Blood collection tubes	A tube made from sterile glass or plastic which is used for collecting blood samples. The blood collection tube may either contain an additive that accelerates the clotting of the blood or an anticoagulant which prevents the blood from clotting	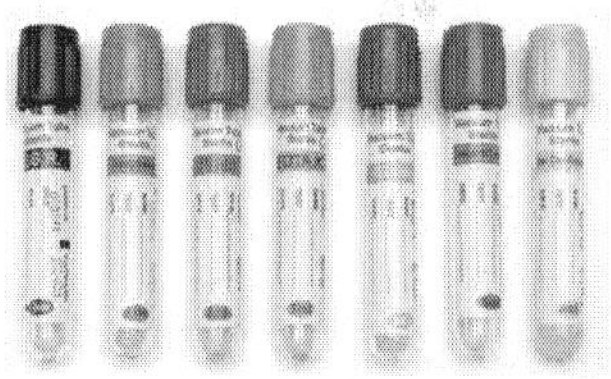
Multiplate photometer	An instrument that measures the intensity/brightness of light produced by an unknown source in relation to a known/standard source	

5.5. STERILITY EVALUATION LABORATORY

A sterility evaluation laboratory focuses on testing for a full range of human pathogens. These include aerobic and anaerobic bacteria; yeasts and molds; blood, tissue, enteric parasites and mycobacteria. Antibiotic susceptibility testing may be performed on bacteria while antifungal susceptibility testing may be done on yeast and molds. These labs may provide support to epidemiology and infection control programs, may assist with surveillance programs whereby the detect and monitor cultures associated outbreaks as well as support control and prevention of healthcare-associated infections (Table 5.7).

Table 5.7. Type of Microbiology Sterility Tests That Can Be Carried Out in a Microbiology/Sterility Evaluation Laboratory

Type of Testing	Test Type		Test Standard
Microbiology laboratory	Microbiology sterility testing	As per USP without antimicrobial activity	USP 71
		As per USP with antimicrobial activity	USP 71
		As per ISO-sterilization validation-without	Approved protocol based on ISO 11737
		As per ISO-for sample containing antimicrobial property	USP 71
	In vitro bacterial reverse mutation assay (Ames test)	5 strains, single concentration, one extractant	ISO 10993-3
		5 strains, single concentration, two extractants	
		5 strains, five concentrations, two extractants	
	Microbiological testing-general	Anti-Microbial Activity testing: Agar diffusion for two strains/parallel streak method	Approved protocol

The basic equipment required to set up a basic microbiology/sterility evaluation laboratory are highlighted in Table 5.8.

Table 5.8. Basic Equipment Required to Set Up a Microbiology/Sterility Evaluation Laboratory

Equipment	**Definition**	**Example**
Laminar flow	A laminar flow cabinet or tissue culture hood is a carefully enclosed bench designed to prevent contamination. Air is drawn through a HEPA filter and blown towards the user. The cabinet is usually made of stainless steel with no gaps or joints where spores might collect.	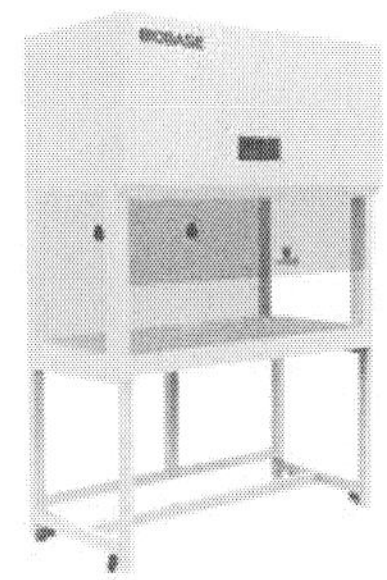
Incubator	An incubator is a heated insulated box that maintains optimal temperature, humidity, gaseous content of the atmosphere inside. It is used to grow and maintain microbiological/ cell cultures.	
Spectrophotometer (UV based)	An instrument for measuring the intensity of wavelengths in a spectrum of light, transmitted by a substance or a solution.	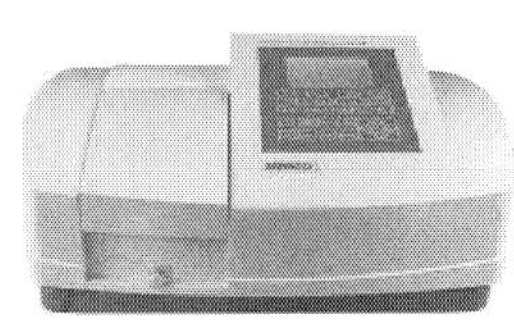

Binocular microscope	A microscope is an instrument used to see objects that are microscopic to the naked eye. A binocular microscope has two eye pieces.	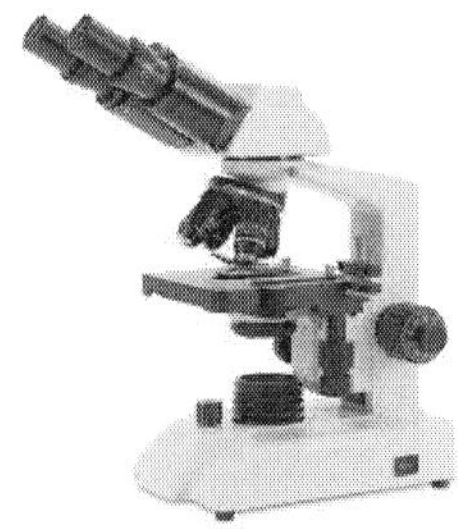
Autoclave	An autoclave is a device used to sterilize equipment and supplies by subjecting them to high pressure saturated steam at 121°C for around 15–20 minutes.	
Sterilizer (Ethylene oxide)	Ethylene Oxide (EtO) sterilization is mainly used to sterilize medical and pharmaceutical products that cannot support conventional high temperature steam sterilization. This is commonly used for devices that incorporate electronic components, plastic packaging or plastic containers.	
Pipettes	A pipette is a laboratory tool commonly used to transport a measured volume of liquid.	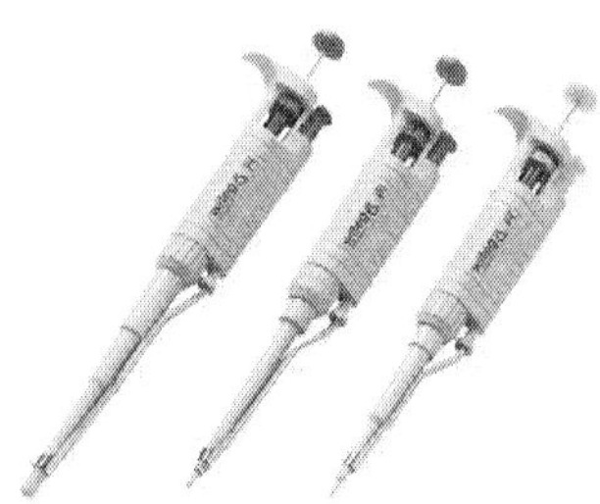

Laboratory water bath	A water bath is a piece of equipment used to heat materials gently and gradually to fixed temperatures, or to keep materials warm over a period of time.	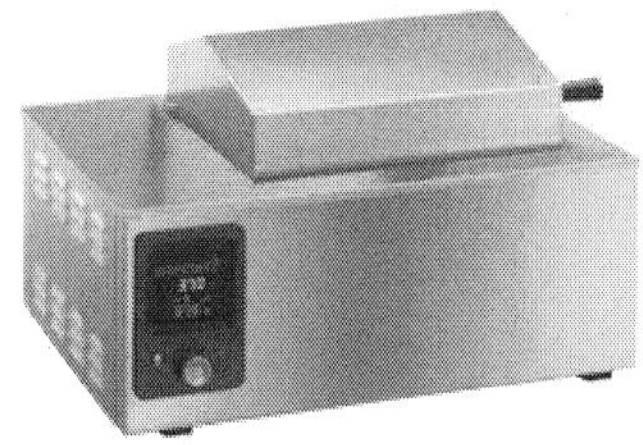
Cryogenic storage vessel	Cryogenic storage vessels are specialized types of vacuum flask used for storing cryogens (such as liquid nitrogen or liquid helium), whose boiling points are much lower than room temperature.	
Basic laboratory glassware/plastic-ware	Laboratory Glassware forms a basic part of scientific laboratories. Ideal laboratory glassware must comprise of borosilicate glass with a low coefficient of expansion, for resistance to heat, and a very high resistance to chemical attack.	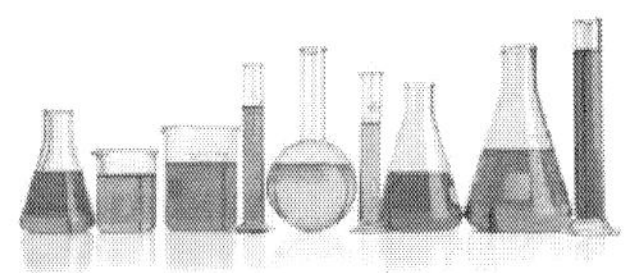
Ultraviolet lamp	Ultraviolet radiation is widely used for killing bacteria or producing fluorescence.	
Air monitoring system	A suitable air monitoring system should be in place which allows for laboratory air temp and humidity to be monitored.	

thermometer	A laboratory thermometer is used to measure temperatures with high accuracy. It can be partially or fully immersed in substance being measured.	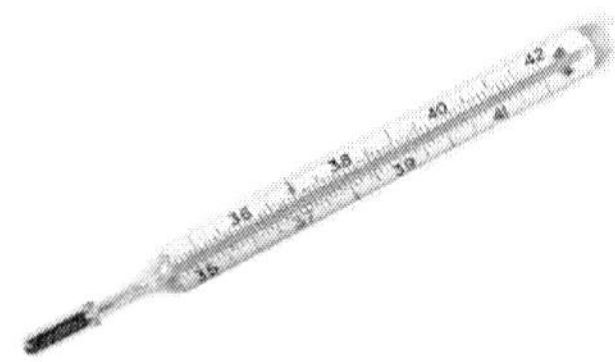
Lab freezer/bio freezer/ultra-low freezer	These are upright or chest freezers designed to protect biological samples by storing at -70/-80 degrees Celsius	
Heating block	A dry block that holds individual containers such as test tubes, micro tubes, or PCR tubes which allows for consistent heating. This is done on a hot plate and allows for the temperature to be controlled.	
Vortex	Simple device used to mix small vials of liquid	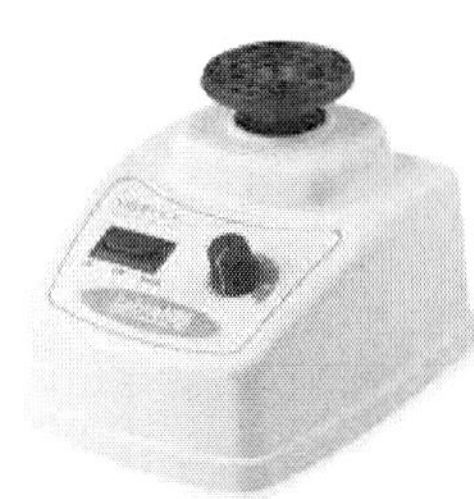

Laboratory use refrigerator	Laboratory grade refrigerator used to keep reagents, chemicals, and media cool	
Hot air oven/drying oven	This is used for the rapid drying of glassware. It can also be used for temperature testing	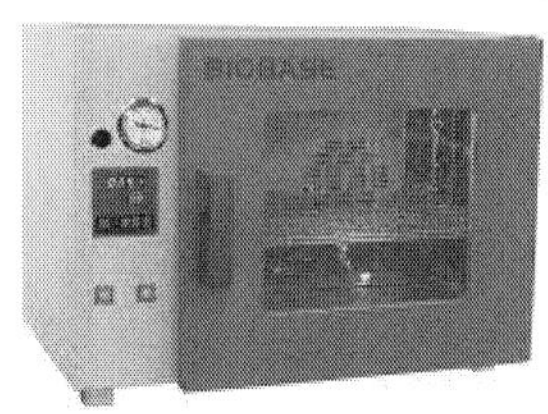
Weighing balance	A device used to measure weight or mass	
pH meter	An electric device used to measure hydrogen-ion activity (acidity or alkalinity) in a solution	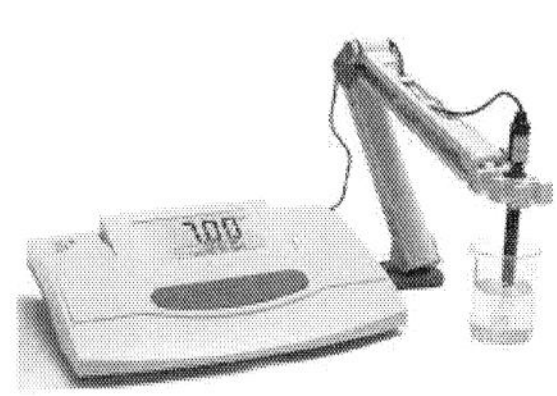

Conductive meter	An instrument that measures the electrical conductivity in a solution	
Microwave	Laboratory grade microwave used to heat agarose, agar, and other laboratory mediums	

5.6. HISTOPATHOLOGY EVALUATION LABORATORY

One of the important measures of hemocompatibility is the hemolysis test, which measures the ability of a material or material extract to cause rupturing of red blood cells (RBC). Hemolysis testing should be performed on all materials that directly come into contact with the bloodstream or entering the circulatory system. A list of histopathological evaluation tests and their relevant standards are mentioned in Table 5.9.

Table 5.9. List of Histopathology Tests

Type of Testing	**Test Type**	**Test Standard**
Histopathology laboratory	Gross and histopathological evaluation-soft tissue	ISO 10993-6
	Gross and histopathological evaluation-hard tissue	ISO 10993-6

The basic equipment required to set up a basic histopathology evaluation laboratory are highlighted in Table 5.10.

Table 5.10. Basic Equipment Required to Set Up a Basic Histopathology Evaluation Laboratory

Section of Laboratory	Equipment	Definition	Example
Processing area	Laminar flow	A laminar flow cabinet or tissue culture hood is a carefully enclosed bench designed to prevent contamination. Air is drawn through a HEPA filter and blown towards the user. The cabinet is usually made of stainless steel with no gaps or joints where spores might collect.	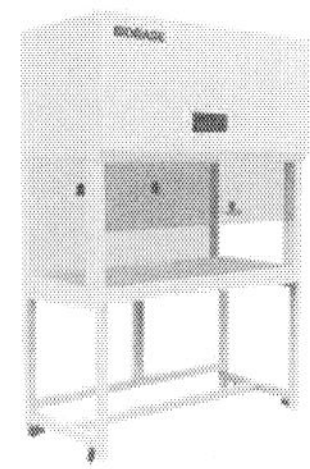
	Macro digital imaging system	A system to capture high resolution images and video files	
Staining area	Auto Stainer	This instrument automates the staining process and allows for high throughput	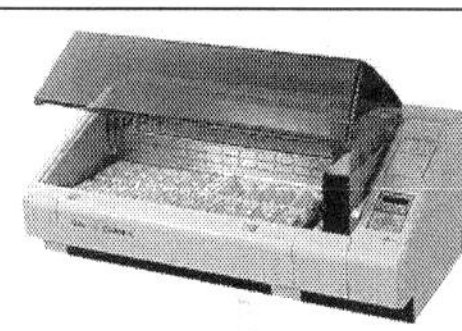
	Cover slipper	This instrument automates the task of adding cover slips to microscope slides	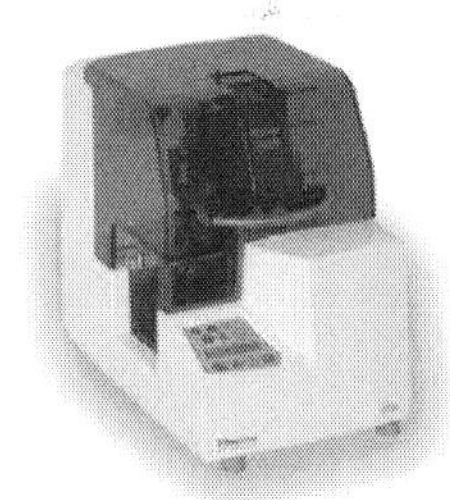
Sectioning area-soft tissue	Fully automated rotary microtome sectioning equipment	This instrument is used for the cutting or sectioning of tissues	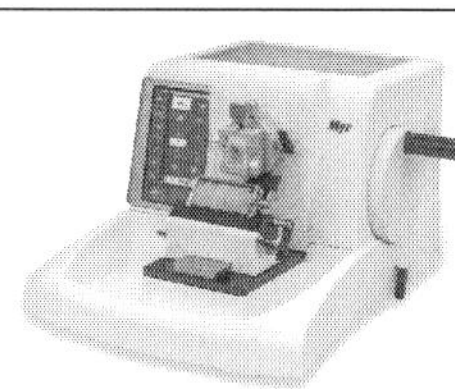

	Laboratory water bath	A water bath is a piece of equipment used to heat materials gently and gradually to fixed temperatures, or to keep materials warm over a period of time.	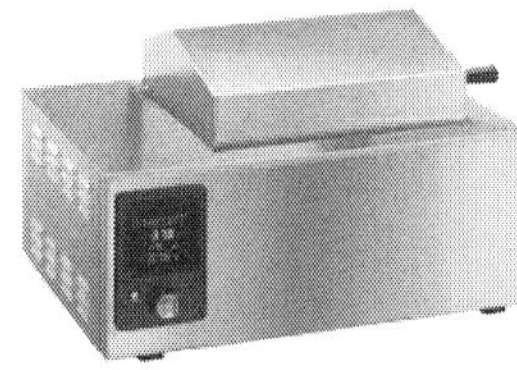
Sectioning area-hard tissue	Tissue grinder	A tissue grinder is used to completely homogenize small amounts of tissue	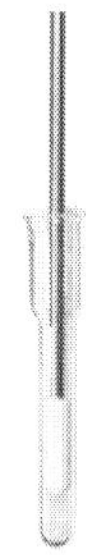
	Various types of microscope	A microscope is an instrument used to see objects that are microscopic to the naked eye. A binocular microscope has two eye pieces.	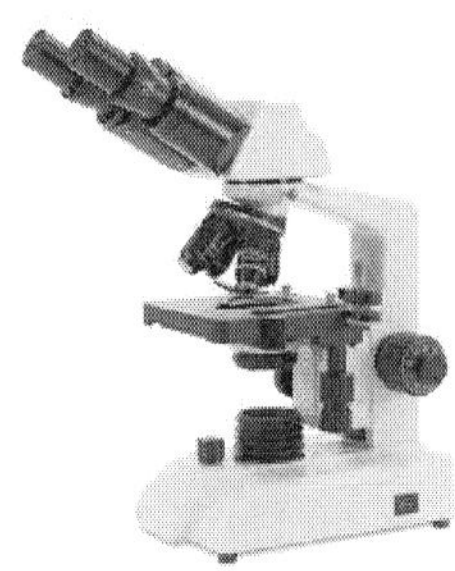
	Pipettes	A pipette is a laboratory tool commonly used to transport a measured volume of liquid.	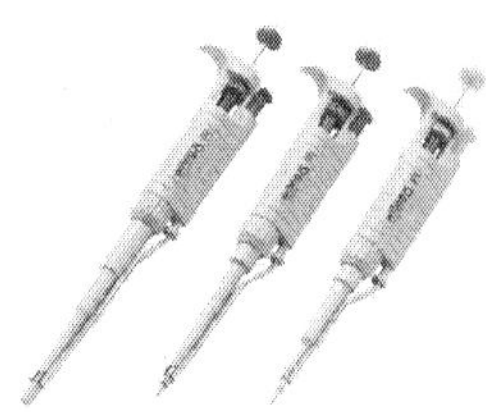
	Cryogenic storage vessel	Cryogenic storage vessels are specialized types of vacuum flask used for storing cryogens (such as liquid nitrogen or liquid helium), whose boiling points are much lower than room temperature.	

	Basic laboratory glassware/plasticware	Laboratory Glassware forms a basic part of scientific laboratories. Ideal laboratory glassware must comprise of borosilicate glass with a low coefficient of expansion, for resistance to heat, and a very high resistance to chemical attack.	
	Ultraviolet lamp	Ultraviolet radiation is widely used for killing bacteria or producing fluorescence.	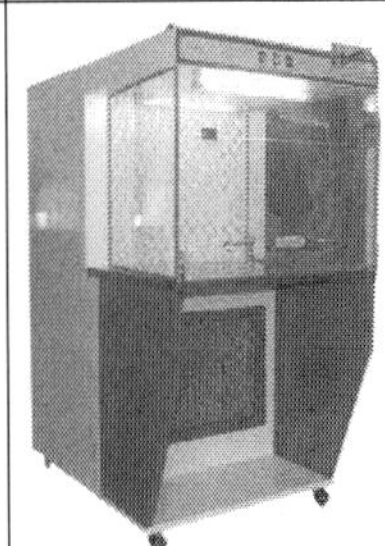
	thermometer	A laboratory thermometer is used to measure temperatures with high accuracy. It can be partially or fully immersed in substance being measured.	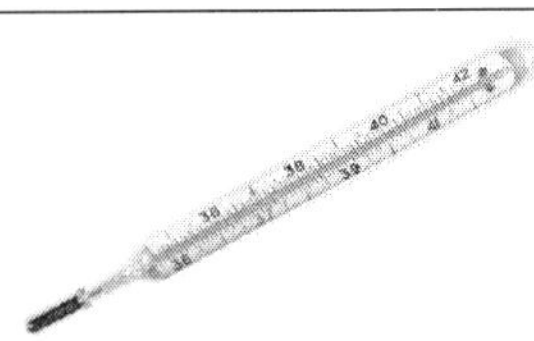
	General laboratory instruments	This encompasses the various tools and equipment used by scientists on a day to day basis. This includes Bunsen burners, burettes, petri dishes, graduated cylinders, laboratory scales and so on	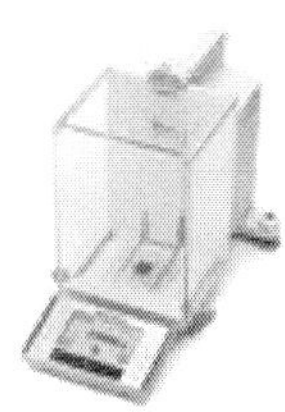

	Chemical testing instruments	Chemical testing equipment includes those pieces of equipment used to determine, characterize, and quantify chemical components. This can include spectroscopy systems, gas chromatography and liquid chromatography systems, process analyzers, thermal analyzers, calorimeters, and so on	

5.7. PHYSICOCHEMICAL EVALUATION LABORATORY

A physiochemical evaluation laboratory focuses on polymers, polymeric formulations, composites, and devices suitable for different biomedical applications. Generally, these types of laboratories are equipped with facilities for compounding/ mixing, molding, extrusion, and electro spinning facilities. An overview of the types of tests possible in a physicochemical evaluation laboratory is highlighted 5.11.

Table 5.11. List of Physicochemical Tests

Type of Testing	Test Type	Test Standard
Physicochemical evaluation laboratory	X-ray Diffraction Spectrum	Approved protocols
	Vicker's Micro Hardness Testing	Approved protocols
	Mechanical Testing	Approved protocols
	Dynamic Mechanical Analysis	Approved protocols
	Mechanical Testing Using UTM	Approved protocols
	FTIR Spectroscopy	Approved protocols
	FT Raman Spectroscopy	Approved protocols
	Thermo cycler	Approved protocols
	Micro CT Imaging	Approved protocols
	Water sorption and Solubility	As per ISO

	Thermal Analysis	DSC in O_2/Nitrogen Atmosphere	ASTM E1256-03
		DTA (RT to 1400 degree C)	Approved protocols
		TGA (RT to 1400 degree C)	ASTM 1131-03
	GPC/HPLC/GC		ASTM D 5296/ANSI
	FTIR Spectroscopy (ATR/KBr)		Approved protocols
	UV-VISIBLE light Spectroscopy		Approved protocols
	Transmission Electron Microscopy (TEM) Analysis		Approved protocols
	SEM Analysis		Approved protocols
	Confocal microscopy		Approved protocols

The basic equipment required to set up a basic physicochemical evaluation laboratory are highlighted in Table 5.12.

Table 5.12. List of Equipment Required to Set Up a Basic Physicochemical Evaluation Laboratory

Equipment	**Definition**	**Example**
FTIP spectroscope	This piece of equipment is used to identify organic, polymeric, and inorganic materials	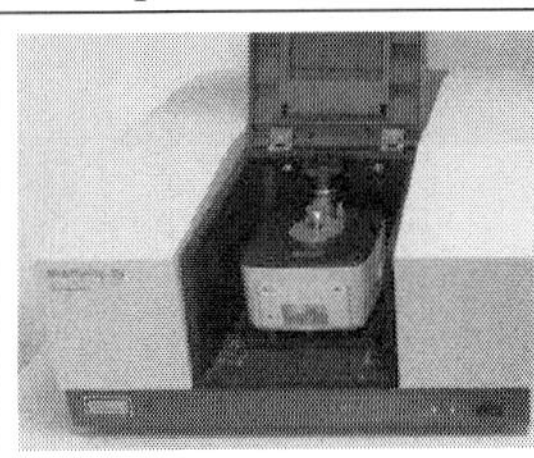

Thermo cycler	Used to amplify DNA and RNA samples by the polymerase chain reaction	
Micro CT imager	Aids high resolution imaging	
Micro hardness tester	Used to determine material hardness or resistance	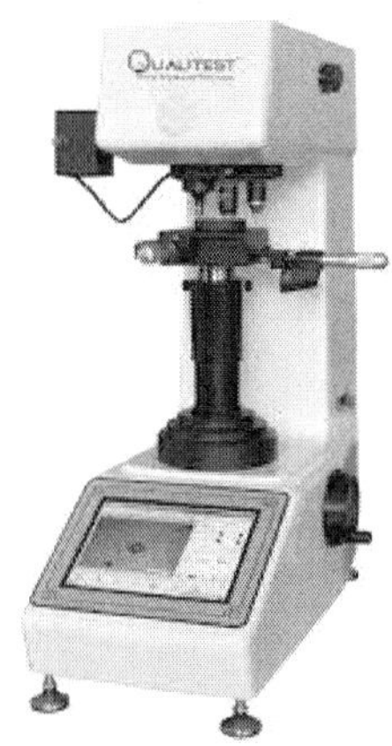
x-ray diffraction spectrum analyzer	Used to obtain information about the structure of crystalline material	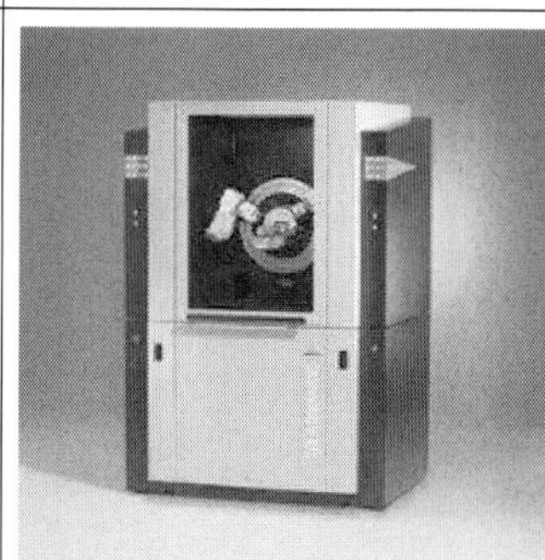

High pressure liquid chromatography	Used to separate, identify, and quantify each component in a mixture	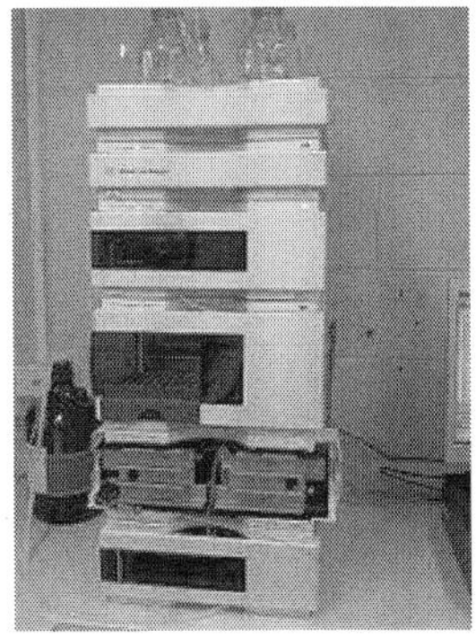
Gas chromatography	An instrument used for the identification and compounds in a mixture	
UV spectrophotometer	An instrument for measuring the intensity of wavelengths in a spectrum of light, transmitted by a substance or a solution.	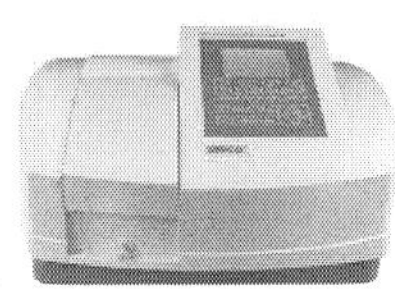
Transmission electron microscope	Type of electron microscope used to view very small samples	
Confocal microscope	A specific fluorescent microscope that allows 3D imaging of a sample with good resolution	

5.8. ACCELERATED AGING LABORATORY

The evaluation of actual shelf life or the long-term operating life of a product or system under real time conditions would ideally require years of time. Hence the need to test such claims under accelerated aging conditions. Basic equipment required for an accelerated aging laboratory is listed in Table 5.13.

Table 5.13. List of Equipment Required to Set Up an Accelerated Aging Laboratory

Equipment	Definition	Example
Torsion testers	Provides dependable torsion, axial-torsion, and low torque testing capabilities	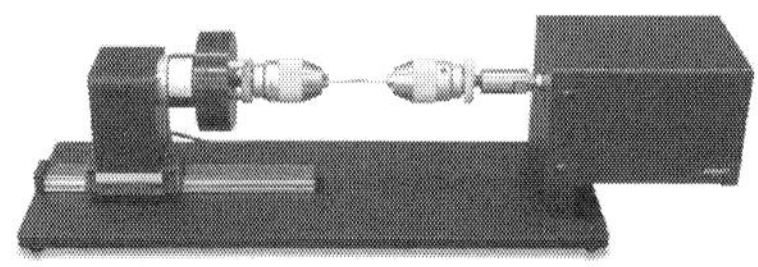
Instron	Single column system performing a peel test	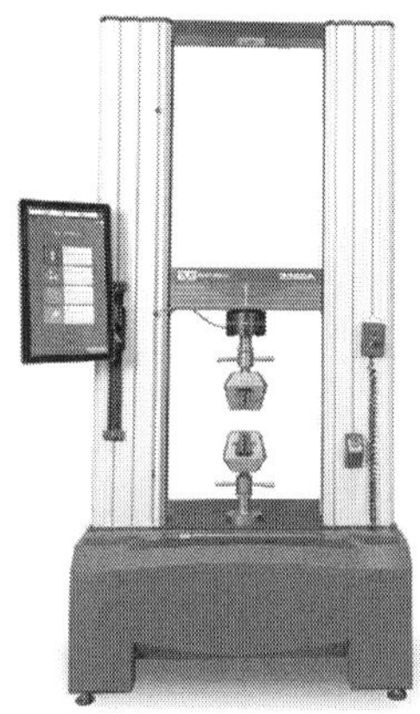
Servo-hydraulic testing machine	This instrument is used to examine the physical strength of biological materials	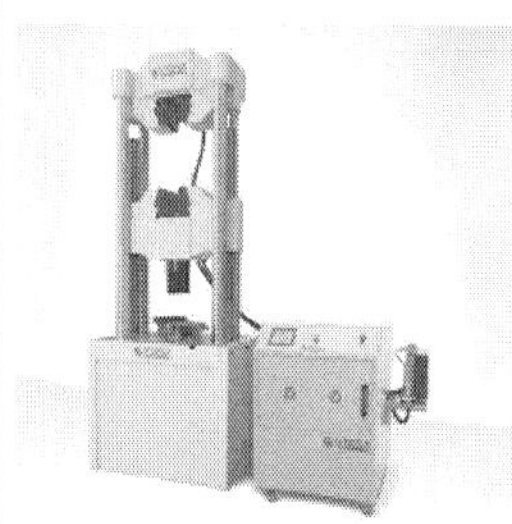

Torque testing machine	This instrument provides moment or rotational movement to an object. It uses an optical rotary encoder to measure rotary motions. It used to evaluate surgical equipment and procedures as well as bone repair methods and materials. This equipment is used to determine failure torque, energy, and stiffness of these materials	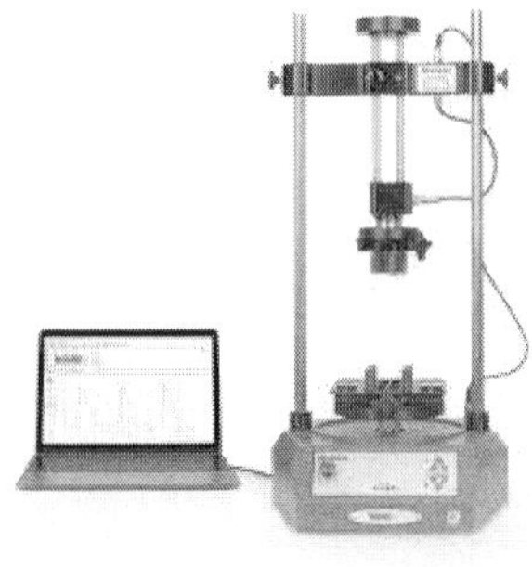

5.9. PACKAGING VALIDATION LABORATORY

A packaging validation laboratory provides package testing services that evaluate the strength and integrity characteristics of a packaging system before and after simulating anticipated distribution and storage conditions that the system may undergo. This testing is often done in compliance to ASTM, ISO, ISTA, and other accepted industry standards. A packaging validation laboratory carries out a variety of tests such as vibration, physical shock, thermal shock, friction, flow rate, force to operate, leakage, and compression testing.

5.10. HOW TO SET UP A GENERAL LABORATORY

Setting up any kind of testing laboratory requires not only sourcing of various laboratory instruments and equipment but an evaluation of the infrastructural layout and consideration of the human resources portion as well.

5.10.1. Human Resources

The human resources sector of any laboratory set up plays a vital role in its success. The number of personnel needed as well as their educational backgrounds and experience all need to be taken into consideration and will be determined based on the type of testing which the lab is offering. In addition, this is also influenced by the methods chosen and the expected sample output. The very first part of setting up a laboratory is to create an organized structural plan for the test laboratory and define the activities or type of testing that would take place. An overview is provided in Figure 5.1.

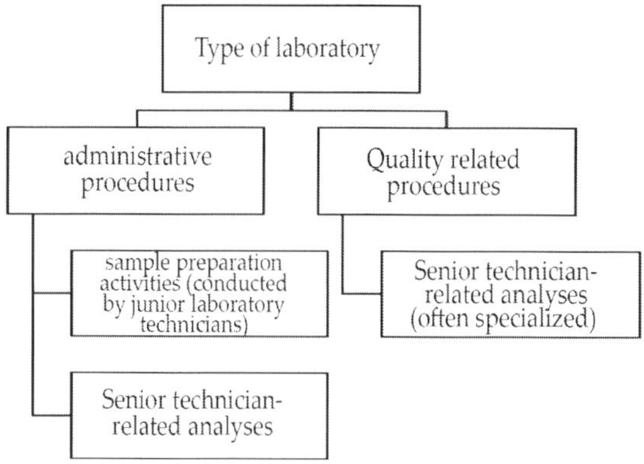

Figure 5.1. A simplified laboratory structural plan.

Ideally, a senior scientist will be responsible for each area of testing with about two or three laboratory technicians. These technicians tend to rotate between activities being carried out which allows for overall knowledge to be gained of the working requirements for each area of the laboratory. In addition, often routine testing can become monotonous and sometimes boring for laboratory technicians who are carrying out the same testing every day. For this reason, it becomes important for senior management to find ways for continuous development of their staff as well as ways to keep the job interesting and challenging for staff. Typical positions within a laboratory include the following.

5.10.2. Laboratory Manager

A laboratory manager is responsible for the whole laboratory as well as the development of its strategic plan. Often much of his or her duties involve external communication with clients and potential clients. In addition, the laboratory manager will usually assume full responsibility for results reported to clients. Any form of management system implemented must be overseen by the laboratory manager so as to ensure reliable data is produced and that the reporting of this data is thoroughly checked prior to release.

5.10.3. Quality Assurance Manager

The Quality Assurance or QA Manager is responsible for quality assurance within the laboratory. Ideally the QA Manager should have an independent position so as not to develop any bias which may influence his or her decision where quality, safety, and functionality of the product is concerned.

5.10.4. Senior Laboratory Technician

A Senior Laboratory Technician is responsible for the daily organization of analytical processes within the lab. He or she is also tasked with ensuring that daily and weekly deadlines within their perspective divisions are met; that quality control for each batch of testing meets requirements and is recorded properly; that staff training is up-to-date and that there are sufficient staff to meet the workload requirements. In addition to the abovementioned, they are also responsible for maintaining stocks of the necessary chemicals and consumables. He or she should have strong and good communication with the Laboratory Manager and QA Manager to ensure ordering and delivery of consumables/ chemicals is done prior to stocks running low as well as for trouble-shooting and maintenance of equipment.

5.10.5. Junior Laboratory Technician

A Junior laboratory technician is generally responsible for carrying out any analytical work in line with Standard Operating Procedures (SOPs), under the direction of the Senior Technician.

From the outset, a laboratory should have all expertise needed to perform all the testing it offers. This means recruiting senior technicians with the background needed to ensure appropriate test methods are carried out. In addition, relevant Standard Operating procedures (SOPs) and Work Instructions should be prepared, a suitable quality control program devised as well as suitable use of RM needed should be put in place to ensure that the laboratory consistently produces results of high accuracy. The procedures devised should ensure that the laboratory can prioritize and organize its workload as well as guarantee the quality of results produced. These should include:

- Acceptance criteria for samples;
- Sample preparation;
- Description of test methods;
- Description of how to validate results;
- Quality control;
- Maintenance and calibration records of equipment;
- HR procedures which link to job descriptions, including

responsibilities and continuing competence of individual technical staff;

- Training records of technical staff, covering which methods they can perform, level of training, whether they can perform a method independently or under supervision, and their ability to train others, etc;
- Traceability and storage of raw data; and
- Cleaning procedures for the laboratory.

5.10.6. Controls within a Laboratory

The highest level of control is often directed at the source. From an occupational health perspective, this may involve immunization of workers who may come into direct contact with infected samples. This includes controls such as vaccines, proper ventilation, needleless systems, improved safety when handling sharps, biological safety cabinets and effective biological waste containment.

Common health and safety issues encountered in a lab include:

- Handling of toxic, corrosive, and flammable chemical hazards;
- Exposure to compressed gases;
- Exposure to blood, body fluids, and tissues which may contain infectious agents;
- Needle-related injuries;
- fires from flammable materials and electronic equipment;
- Physical hazards such as radioactive materials, ultraviolet light sterilizers and lasers;
- Handling of cryogenic (ultra-cold) materials;
- Standing and working in awkward positions or for long periods of time, creating risks of back and arm injuries;
- Repetitive motion injuries;
- Electrical hazards from electrical equipment and instruments;
- Hazards associated from spilled liquids and congested work areas;
- Cuts and lacerations from broken glass; and
- Burns and scalds from hot equipment.

5.10.7. Layout of a Laboratory

Laboratory design is multi-faceted and plays a huge part in the overall research process and should be done in such a way that it promotes efficient scientific testing. Even the smallest error in design can lead to delays in testing with an increase in possibility for mistakes, accidents, and compromise the overall output. Depending on the type of testing done, it is important to consider if there needs to be separation of facilities or test sites in separate labs or buildings. This is ideally the most appropriate arrangement for the minimization of cross-contamination. If, however, this is not possible, then the lab layout and design should consider an alternative approach which allows for the use of certain limited and common facilities in one particular "clean" area with office facilities central to each activity so as to further avoid cross contamination. Often the laboratory manager is faced with the task of having to modify existing accommodation due to budgetary constraints. This often involves a compromise in the initial design however there needs to be both soft and hard limits when taking this compromise into consideration. Some activities may be combined while others cannot.

While the internal design is often a matter of preference, the work load, work pattern as well as relevant ISO standards will guide and determine the layout. It is important to ensure the overall design matches the objectives and need of the laboratory itself. Ideally a separate weighing room with a vibration proof balance table and a room with ultra-freezer facilities for storage of samples should be planned for. Partitioning of one of the main rooms is acceptable. Where large numbers of samples are anticipated, the provision of a walk-in fridge should be considered. The need for a chemical store must also be considered and this should be situated outside of the laboratory. Lockable cupboards should be used where possible and solvents should be kept on trays and in metal lockable cupboards.

5.11 ESSENTIAL POINTS TO CONSIDER WHEN DESIGNING A LABORATORY

5.11.1. Pre-Planning

Initial planning should involve all stakeholders and this should be done early and often in the design process. Employing a multidisciplinary approach is critical to ensure different perspectives are gained, clear understanding of workflows and processes as well as identification of critical lab requirements such as utilities, emergency power, lab equipment, ceiling heights and storage requirements. This provides an invaluable foundation to the holistic functionality of the laboratory. Commonly included stakeholders include property owners, laboratory technicians, laboratory managers, facility, and maintenance personnel as well as any other essential member of staff which could include quality, engineering, and supply chain.

5.11.2. Location, Layout, and Access

Establishing the location, layout, and size of the laboratory is also important. Failure to efficiently size the laboratory such that it meets user requirements can result in inadequate bench and equipment storage space, inflexibility or limited growth capabilities as well as limited overall functionality. This often leads to downstream spill over into corridors and common spaces which can result in cross contamination and a variety of safety hazards. If the lab is to be established in a multi-story building then the design should take into consideration if vertical transportation will be needed and if there will be convenient access to things like utility services, waste disposal and relevant instrumentation. Common factors that influence the layout of a room include ease of maintenance, accessibility of equipment, HVAC requirements, and ergonomics. Ideally benches used for admin work and sample preparation should be placed around perimeter walls to improve access and workflow and this should take the height and width of the benches into consideration for comfort and efficiency. The lab design should also ensure a certain degree of flexibility or adaptability is employed as this takes future growth and change into consideration. It is possible that processes, research methods and practices will change over time which in turn means that researchers may perform different tasks or two or more researchers may be present performing different tests. This adaptability allows consideration of comfortable working should changes occur. This also means that the build of the space should be easy to reconfigure and worktops and equipment stations should be able to be relocated to fit the goals of a new project, without the need for bringing in outside workers.

5.11.3. Safety

Ensuring a safe and securely designed laboratory is essential so researchers can focus solely on their tasks, rather than having to be overly vigilant when it comes to emergencies and accidents. Safety features should include the use of bio-safety cabinets, fire protection and detection systems, emergency showers/eye wash stations, and well-marked exits, all of which should be easily accessible. The correct ventilation systems, tailored to researchers' activities, should also be in place. In addition, control of temperature should be employed, not only to ensure a controlled environment within the laboratory but also for user comfort.

5.11.4. External Services

The need for external services and their availability must also be taken into consideration. In some countries, power supplies are prone to disruption and subject to frequent voltage fluctuations. Water supplies may be limited, and those countries where they are not, they may be subjected to low pressure and supplied at varying temperatures depending of the climate within the country. Depending on the type of laboratory, some may use relatively large amounts of electrical equipment and the overall consumption may be high. While some equipment is used periodically, a large amount of equipment is run continuously. This includes, laboratory ovens, refrigerators, freezers, and balances to name a few.

For this reason, the design of a laboratory must ensure the power supply is:

- capable of meeting the full demands of the laboratory; and
- be stable and not subject to voltage fluctuations-if voltage fluctuations are likely then a voltage stabilizer must be installed.

Water supply is also essential as most laboratories tend to require fairly large volumes of water. In addition, this supply should be uninterrupted. The main aspects are the total flow and volume of water involved, the output pressure as well as the degree of cooling required for various equipment used. Usually the outlets leading to the washing-up sink or for the generation of a vacuum should be connected to the open mains supply.

A supply of compressed gases will be needed for the laboratory. This should be available for analytical equipment and also for bench top operations. It is important during the design stage that provision is made for the supply of desired gases. This will depend on the range of gases required as well as the flow rates and overall consumption. In addition, the supply mechanism should also be taken into consideration. Some gases will be provided from cylinders and should therefore be located outside the laboratory for safety reasons. These cylinder type gases will require a feeding line from the supply source to the internal network. The cylinders should be contained in a separate storage area outside the laboratory building and should be shielded from direct sunlight, rain, and other weather or temperature extremes. This storage area should only be accessible to authorized personnel but should be located conveniently for the delivery of cylinders when they need to be replenished. Each cylinder must be fitted with a pressure control gauge with which to set the line pressure. For safety reasons, his gauge must be at the cylinder head and not lower down through the supply system. Gauges must be kept clean and grease-free. The pressure setting and hence the specification for the gauge itself will depend on the requirements of the system and, ultimately, the operating pressure which varies between equipment make and the overall nature/type of instrument. Gases may also be supplied from generators or compressors. Generally, these will fit beside or below laboratory bench units and can be contained within a laboratory complex. Gas purifiers must also be fitted in the lines between the supply and outlet, especially if the gases are to be used in analytical experiments. For example, during gas chromatography. Each gas feed must pass through a moisture filter. The nitrogen supply should pass through an oxygen trap. This is essential to preserve the safe and effective working of analytical equipment at high sensitivity. Purifiers must be changed regularly to prevent saturation.

All piped supplies should run through clean, high-purity tubing. Often copper is the material used. Joints must be carefully welded and solder must not be used, as with time this can produce contaminants. It is recommended that on/off valves be fitted where the supply pipes enter the laboratory so that supplies can be turned off in case of need or emergency. Pipe outlets should be capped for cleanliness whilst awaiting connection. The cost of piping installation varies with the complexity of the distribution network but should generally be included in the construction or conversion costs.

The provision of a vacuum supply is also recommended. Some laboratories use a central pump to generate vacuum with a number of outlets fed by a network of piping. Water-generated vacuum is favored on safety grounds if the water pressure and supply are adequate. In cases where vacuum can only be provided through the use of electrically operated pumps, particular care must be taken to prevent liquids or vapors, especially those of an inflammable nature, passing through the pump.

5.12. KEY CONCEPTS OF A BIOCOMPATIBILITY TESTING LABORATORY

The main purpose of biocompatibility testing is to determine how compatible a device is with a biological system. As we have already established, the purpose of performing biocompatibility testing is to determine the fitness of a device for human use, and to see whether use of the device can have any potentially harmful physiological effects. The overall process of determining the biocompatibility of any medical device involves several stages:

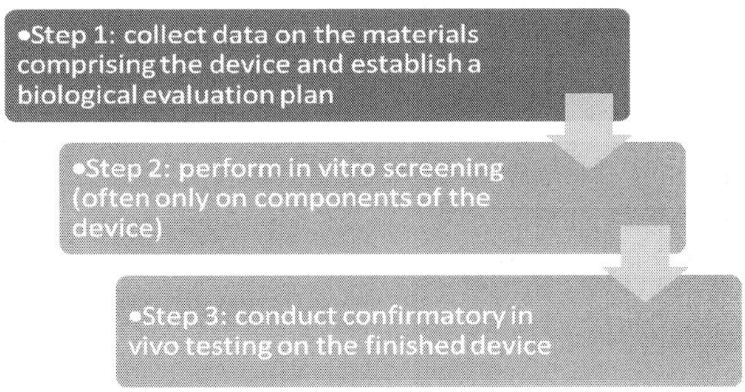

Figure 5.2. Simplified overview of the biocompatibility testing process.

5.12.1. Pre-Test Planning

As we have already determined in previous chapters, there are a range of different biocompatibility tests which are applicable to different types of devices based on the duration of use and type of body contact. In certain cases, it can become confusing when trying to determine which is the appropriate test to choose. Many of these tests are quite costly and depending on the type of device, animal testing may also be involved. For these reasons, it is important for the manufacturer to carry out some research prior to testing in order to properly understand what component materials make up their device and subsequently what type of testing will be needed. Often research into biocompatibility testing of similar devices with an established clinical history may help. In certain cases, existing data may be sufficient to demonstrate biological safety of parts of, on even the entire device, in these cases it may preclude the need to conduct certain types of biocompatibility tests.

Pre-test planning often involves an in-depth look at the intended use of the device. It is important to note and record whether the device is a surface device, an external communicating device, or an implant device, and what tissues the device will contact. For example, implant devices interacting with the blood will require more thorough testing as when compared to a surface device with an expected contact time of only a few days, which will require less testing.

Regulatory submission should also be taken into consideration at this stage. Most biocompatibility testing laboratories offer testing options with and without Good Laboratory Practice (GLP). A manufacturer must determine the main purpose of biocompatibility testing and where the data will be used and/or submitted. If the data is for regulatory purposes, and will be submitted to various regulatory bodies, then testing with GLP is required. In any event, it generally a good idea to conduct biocompatibility testing with GLP to allow for the maximum regulatory flexibility and compliance.

5.12.2. Conducting Tests

Before testing can begin, a biocompatibility testing laboratory will perform material characterization and an analysis of the device's components. This often involves determining if there is a leachable material from the device or related components at an elevated temperature as well as analyzing the leachable extracts for potentially harmful chemicals or cytotoxicity. Testing will usually start with a form of vitro testing and once this has been completed, *in vivo* biological testing can be done but this is based on the device's intended use. The turnaround time for tests can range from anywhere between three weeks to several months but this will depend on the type of test as well as the laboratory conducting the test.

CHAPTER 6

Biological Evaluation Plans and Reports

CONTENTS

6.1. DEVELOPING A BIOLOGICAL EVALUATION PLAN

A Biological Evaluation Plan (BEP) is created to review device materials, identify any potential risks as well as suggest possible evaluations and testing methods to address the risks identified. This is done based on the intended use of the device in relation to the nature of patient contact. The BEP often helps guide manufacturers in finding a suitable approach for testing. It breaks down the components of the device and determines what materials they are composed of. In addition, it further looks at each component of the device and determines individually what type of testing must be done. If a BEP is prepared properly, it will help a manufacturer avoid additional time and expenses. The plan, together with the final overall Biological Evaluation Report (BER) can help a manufacturer to dully understand their device and overall biocompatibility, which is essential when completing regulatory submissions and collating overall product information of the device. These documents become especially useful for US FDA and CE marking submissions.

The ISO 10993-1 standard outlines the importance of having thoroughly analyzed a device before testing can begin. Proper risk management should be done where any biological risks identified are assessed and mitigated. The evaluation of biological properties and determination of biocompatibility hazards as associated risks should be done in well in advance. The information from here should be used to draw up a biocompatibility safety evaluation plan.

The risk management plan should evaluate the physical and chemical properties of the device as well as the device's overall intended use. Any historical clinical use, data recorded from exposure to humans and toxicological and biological data should be included. This information will be used to select the appropriate tests and any additional tests where necessary. If any hazards are identified, then additional research into similar devices and their manufacturing and testing methods will need to be done. Once associated biological risks are identified, then a suitable strategy for a biological safety evaluation plan can be proposed. Once testing has been carried out, a conclusive document that summarizes the biological evaluation performed on the device is drafted. This is known as the biological evaluation report. The BER provides an assessment of the overall safety evaluation of the device. It includes a description of the biocompatibility tests performed and, when necessary, solid rationales to justify why any tests were done and why deviation for standard norms was employed.

Box 20: General Principles and Steps of Biological Evaluation

Identification of applicable standards

Standards and all other relevant documentation applicable for the medical device should be identified. In cases where specific standards exist, it is recommended to use them.

Formulation, description, and manufacturing information

Important components of the medical device should be identified. This includes the chemical name, CAS number if relevant, weight percentage of each chemical component present and function of each chemical component in the device

The characteristics of the finished device should be described according to ISO 10993-19. Important characteristics include connectivity, hard or soft surfaces, hydrophobic or hydrophilic surfaces, ability to absorb or repel proteins, surface friction, stability of treated surfaces, roughness of surfaces, water or solvent absorption properties, size, and shape.

The manufacturing steps of the medical device should be included. Information on processing additives and known or suspected impurities as well as sterilization method used should be included.

Information on the intended use of the device and evidence supporting clinical performance should also be included. Lifetime claims, expiry date and storage conditions should also be included.

Categorization of the medical device: nature and duration of contact

The nature of contact with the body should be defined i.e. determine if the device is surface contacting, external communicating or an implant device.

The duration of contact with the human body should de determined i.e. does the device have limited, prolonged or permanent exposure.

The device should also be defined as either, single-use, multiple use or repeated contact with the human body.

Device material characterization

Material characterization is one of the first steps in the biological evaluation process. This should take into consideration chemical constituents, residual processing additives used in manufacture as well as identification of the presence and nature of leachable substances.

Characterization should identify if potential degradation may occur, as well as what the nature of such degradation products may be. Any biological hazards should also be identified. If devices undergo a change, then the tested product should be representative of the device final version.

Analysis of the results, literature, and research method
Any literature research done to interpret the results obtained should be included and this should take into consideration the categorization of the medical device in terms of nature and duration of contact. Use of historical data held by the manufacturer, if relevant, should also be included.
Leachables which have known toxicological data and verification of the adequate safety margins should be included. If several leachables are present then an analysis of their synergistic potential should be done.
Assessment of the time of release and quantity of each leachable should be done.
The same considerations mentioned for leachable should be done for degradation products.
A protocol should be established for the identification, selection, collation, and review of all available studies and data. The objectives of the literature review should be clearly defined as well as criteria for selecting or excluding such data. A justified conclusion based on the state of the art and analyzed data should be included.
A suitable conclusion should be present which summarizes data from the literature review, historical data held by the manufacturer and any potential hazards. The conclusion should also advise if proceeding to the next stage is recommended or not.

Establishing a biological testing program
After the above-mentioned steps are carried out, the determination of the need for additional biological testing should be done.

Analysis of results
This should involve analysis of historical data held by the manufacturer, if relevant, literature data and results of biological testing.

Feedback to risk management
Finalization of the risk management file should be done. This should include an evaluation on the control of identified potential risks to an acceptable level.

6.2. RE-EVALUATION OF BIOCOMPATIBILITY DATA

Manufacturers should note that once biocompatibility testing is done, it does not mean their device is exempt from future testing of biocompatibility review by regulatory bodies. The data should be re-evaluated in the following instances:

- Changes in the source or specification of materials used to manufacture the product. This includes individual components and additives;

- Changes in the physical configuration of the final device. Common examples of configuration changes include changes in size, geometry, and surface properties;
- Changes in the formulation, processing, and/or primary packaging of the device;
- Changes in the sterilization process;
- Changes in the manufacturing process;
- Changes in the manufacturer's instructions or expectations around storage of the device (shelf-life and/or transportation);
- Changes in the intended use of the product; and
- Any evidence that the product may produce adverse effects when used in humans.

The results of all tests should be summarized in a Biological Evaluation Report (BER) which forms a part of the overall biological evaluation. This report is often submitted to regulatory authorities, who may then, further request the other components of the biological evaluation such as actual BEP, actual test reports from the laboratory, laboratory accreditation certificates and the risk management report.

6.3. TEMPLATE OF A BIOLOGICAL EVALUATION PLAN

Purpose/Objective

The purpose of the BEP should be highlighted here. Relevant standards should be mentioned.

Example: The purpose of this biological evaluation plan is to have a structured program that focuses on the biocompatibility risk associated with the various materials used during the manufacturing process.

Applicable Standards

Any applicable standards relevant to a manufacturers BEP should be highlighted here. This should include relevant parts of the ISO 10993 series.

Scope and Description

This section should provide an overview of which devices and device components are covered by this BEP. The intended use of the medical device should be mentioned.

Example: This plan covers all the materials, additives, and processes used to assemble (product name). The product consists of (list all device components making up the final product).

Categorization of the Device

This section should provide a breakdown of all the device components and details around each component. This should include details of type of steel used, supplier details, sizes, pigments, variant types, additives used, etc.

If the device consists of multiple components, then each component should be analyzed for nature and duration of body contact in line with ISO 10993. This can be done in tabular format for ease of understanding.

Example of a table showing categorization of a device based on ISO 10993-1 and identification of applicable tests:

Material description	Contact duration A- Limited ($\leq$24 h) B- Prolonged (>24 h to 30 days) C- Permanent (>30 days)	Cytotoxicity	Sensitization	Irritation or intracutaneous reactivity	Systemic toxicity acute	Subchronic toxicity (Subacute toxicity)	Genotoxicity	Implantation	Hemocompatibility
Component 1									
Component 2									
Component 3									
Component 4									

Where appropriate, the total surface area of the device which is in contact with the body should be defined. The manufacturing process can de described here and all known/possible biological hazards can be identified.

Description of tested items

A rationale for the selection of the sample to be tested should be provided. Tests should be assigned based on the appropriate biological effects. Chemical analyses for the evaluation of extractables should be identified. Distinction between *in vitro* and *in vivo* testing should be made.

A description of test samples used should be given. This should include as much detail as possible, including LOT/ REF No., manufacturing date and so on. A statement on the sterile state of the same should be given and a rationale that sterilization has no influence on the biocompatibility of the final device. A rationale of the test sample or related variant chosen should be given, for example if a worst-case scenario is chosen and why.

Overview of biological effects considered

This section should provide an overview of the biological effects that will be considered to determine the testing required. Reference should be made to:

- Verification of the availability of previous biocompatibility data for all components of the device tested;
- Identification of any additives added;
- Verification that all additives used during the manufacturing process were part of the biological testing;
- Analysis of product bioburden levels on component parts and packaging materials; and
- Analysis of any residual on the finished product.

Justification for biological effects not considered

This section should provide an explanation for which tests are chosen or not and why. If a material has a documented safe history of use and if it warrants exclusion of certain testing, then a detailed explanation should be provided. Any relevant preclinical studies and clinical experience shall be included. Data available from suppliers may be included to support the justification but full transferability of data must be shown.

Conclusion/Summary of overall evaluation

This section should provide a review of all available toxicity data for each material/component of the device in contact with the body. The overall evaluation should make reference to any additives, catalysts, solvents, sterilization agents, processing chemicals, degradation products, etc. used. A critical overall evaluation of the literature should be done and a compilation of tests to be performed shall be included.

This section should include as much detail as possible. It should make reference to:

- The test sample and if this was a part of the device or whole device;
- Characteristics and specifications such as polymer type, trade name;
- Status of the test sample (i.e., is it the final product and is it sterile);
- Type of body contact involved;
- Contact duration;
- Applicable standards;
- Extract preparation (i.e., medium used, surface/mass-to-volume ratio, temperature, time)
- Overall residual risk-benefit evaluation; and
- Reference to the biological evaluation report.

6.4. TEMPLATE OF A BIOLOGICAL EVALUATION REPORT

Purpose/Objective

The purpose of the biological evaluation report should be highlighted here.

Example: The purpose of this biological evaluation report is to evaluate the biocompatibility of (product name) and all relevant components and additives used during the manufacturing process.

Applicable standards

Any applicable standards relevant to a manufacturers biological evaluation should be highlighted here. This should include relevant parts of the ISO 10993 series.

Scope and Description

This section should provide an overview of which devices and device components were subject to the biological evaluation. The intended use of the medical device should be mentioned.

Biological Evaluation approach

An overview of the biological evaluation approach adopted should be provided here. This section should link to the BEP and ISO 10993-1: Biological Evaluation of Medical Devices.

Description of tested items

A rationale for the selection of the sample to be tested should be provided. Tests should be assigned based on the appropriate biological effects. A breakdown of all device components subjected to testing should be provided as well as any relevant details. This includes type of steel used, supplier details, sizes, pigments, variant types, additives used, etc

Further details to be included are LOT/ REF No., manufacturing date, etc. pertaining to the batch sent for testing.

Analysis of testing results

This section should provide an overview of all tests done, results obtained and if it was a pass or fail. Reference should be made to the relevant test reports. A simple tabular format can be used:

CHAPTER 7

Easy Review Questions

1. **What is the first thing to be aware of when considering biocompatibility?**

   If/when you are considering biocompatibility testing, one of the first things to become familiar with is the UISO 10993 series of standards. ISO 10993-1 standard forms the framework for biocompatibility testing and provides guidance for the biological evaluation of devices within a risk management process. It provides a detailed explanation on how to plan biological assessments of medical devices. Once you have an idea of which tests are applicable to your particular device, use the remainder of the ISO 10993 standards in the series to better understand the applicable tests.

2. **How long does biocompatibility testing take?**

   The "big three" biocompatibility tests, also known as cytotoxicity, irritation, and sensitization tests are usually required for most low risk devices. These tests usually take anywhere between 5–8 weeks. If additional testing is required, such as systemic/subchronic toxicity or implantation studies, then testing timeframes can increase drastically to anywhere between 26–28 weeks. Depending on the laboratory conducting the test, and if a manufacture requests testing with good laboratory practices (GLP), testing can be longer.

3. **How much does biocompatibility testing cost?**

   The cost of biocompatibility testing varies depending on the type of device and overall intended use. This in turn will determine what type of testing is required. It should be noted that biocompatibility determination can become the most expensive part of the development process with costs ranging anywhere from $10,000 to over $100 000 depending on the types of testing program being followed.

4. **What should manufacturers know before starting biocompatibility discussions?**

   Manufacturers/ developers should be able to clearly identify what materials/individual components are used to construct their device as well as what the overall intended use of the device is.

5. **How does regulatory requirements impact biocompatibility testing?**

   It is very important for manufacturers to understand what the purpose of completing biocompatibility testing is. Many laboratories offer a testing with GLP and without GLP option. If the data obtained is intended to be submitted to regulatory authorities for territory specific approval then manufacturers should always select the testing with GLP option. Manufacturers should also note that why many individual materials may have been proven safe by other manufacturers, literature or even historical market approval, many regulatory bodies require manufacturers to be able to prove biocompatibility on production-equivalent/finished devices.

6. **How do you choose a suitable biocompatibility testing facility?**

   Manufacturers should choose a laboratory that is skilled in biocompatibility testing medical devices specifically. While there are many laboratories out there, it is important to screen these laboratories to ensure they have the correct certification and that they can conduct testing as per relevant regulatory requirements. It's a good idea to set up a preliminary call that allows for introductory discussions whereby a manufacturer can assess the level of knowledge and support a laboratory can provide.

A simple laboratory comparison template for manufacturers to use is presented below:

	Laboratory 1	Laboratory 2	Laboratory 3
Type of test			
Cost of test			
Duration of test			

Screening questions			
Sample handling			
How can we be sure that our samples have reached you intact and still sterile before you begin testing?			
How will our samples be handled in your test facility and during the test assay (i.e., maintenance of sterility)?			
Experimental			
Our samples have a fairly small/large Surface Area/Vol ratio. How will you ensure that the entire sample is tested correctly?			
What dilutions will you be running in the assay?			
How many replicates will the lab run per sample?			
Can you tell us more about the positive controls being used?			
Can you tell us more about the negative controls being used?			
How will you determine what is a pass and/or fail?			
Would a repeat run with more replicates be suggested if the test fails or if there are too many outliers?			

As we are in the process of pursing CE marking with the intent for these devices to enter the European market, do you recommend testing with GLP or without GLP?			

Bibliography

1. 21 CFR Part 58. Code of federal regulations title 21, Chapter 1, Subchapter A, Part 58; Good Laboratory Practice for Nonclinical Laboratory Studies.
2. A Disaster Relief Supply Management System (SUMA), (1996). Pan American Health Organization/WHO Regional Office for the Americas, Washington.
3. AAMI Standards and Recommended Practices, Volume 4: *Biological Evaluation of Medical Devices*, which includes AAMI/ANSI/ISO Standard 10993. (Annex B of 10993-1 is an extensive bibliography of U.S. and international reference documents.)
4. All content copied from the PMI Science Website is copyright Philip Morris Products SA-https://pmiscience.com (accessed on 29 May 2020).
5. Andersen, K. E., & Maibach, H. I., (1985). Guinea pig sensitization assays: An overview. In: *Contact Allergy Predictive Tests in Guinea Pigs* (pp. 59–106). Karger New York.
6. Andersen, K. E., Volund, A., & Frankild, S., (1995). The guinea pig maximization test with a multiple dose design. *Acta. Derm. Venereol., 75*, 463–469.
7. Anonymous. (1995a). *Guidelines for Basic Biological Tests of Medical Materials and Devices (Japan).* Tokyo, Japan: National Institute of Health Sciences.
8. Application of risk management to medical devices, (2007). International Organization for Standardization (ISO 14971:2007 Standard). https://www.iso.org/obp/ui/#iso:std:iso:14971:ed-2:v2:en (accessed on 29 May 2020).

9. ASTM F-748-98. (1998) *Practice for Selecting Generic Biological Test Methods for Materials and Devices.*
10. ASTM. (1995). F 748-95 standard practice for selecting generic biological test methods for materials and devices. *Annual Book of ASTM Standards, Section 13, Medical Devices and Services* (pp.193–197). West Conshohocken, PA: ASTM.
11. Buehler, E. V., (1964). A new method for detecting potential sensitizers using the guinea pig. *Toxicol. Appl. Pharmacol., 6*, 341.
12. Buehler, E. V., (1965). Delayed contact hypersensitivity in the guinea pig. *Arch Dermatol, 91*, 171–177.
13. Buehler, E. V., (1994). Occlusive patch method for skin sensitization in guinea pigs: the Buehler method. *Food Chem. Toxicol., 32*, 97– 101.
14. Buehler, E. V., (1994). Protocol. *Delayed Contact Hypersensitivity Study in Guinea Pigs (Buehler technique)*. Cincinnati, OH: Hill Top Biolabs.
15. Casas, J. W., Lewerenz, G. M., Rankin, E. A., et al., (2013). *In vitro* human skin irritation test for evaluation of medical device extracts. *Toxicol. In Vitro, 27*, 2175–2183. doi: 10.1016/j. tiv.2013.08.006.
16. Center for Devices and Radiological Health, US FDA, (1987). *Tripartite bio-Compatibility Guidance for Medical Devices*. Blue Book Memorandum G87-1. Rockville, MD: US Food and Drug Administration. Available from: http:/www.fda.gov/cdrh/g87-1.html (accessed on 29 May 2020).
17. Chang, T., (1997). *Medical Device Donations to Developing Countries: Things to Consider.* Poster presentation at the 32nd AAMI annual meeting, Washington, D.C., USA.
18. Clinical investigations, (2010). *Global Harmonization Task Force.* GHTF/SG5/N3:2010 http://www.imdrf.org/docs/ghtf/final/sg5/technical-docs/ghtf-sg5-n3-clinical-investigations-100212.pdf (accessed on 29 May 2020).
19. Coleman, K., Dai, X., Deng, X., et al., (2012). Medical device biocompatibility evaluation: An industry perspective. In: Boutrand, J., (ed.), *Biocompatibility and Performance of Medical Devices*. Cambridge, UK: Woodhead Publishing Ltd. doi: 10.1533/9780857096456.2.201.
20. Development of medical device policies, (2011). Geneva: World Health Organization. WHO Medical Device Technical Series. http://apps.who.int/medicinedocs/documents/s21559en/s21559en.pdf (accessed

on 29 May 2020).

21. EC, (1992). *EC Directive 92/69/EEC, Official Journal of the European Communities, L 383, Annex, Part B: Methods for the Determination of Toxicity, B.6.* Acute toxicity, skin sensitization.
22. EC, (1993). *EC Standards, EC Directive 93/21/EC, Official Journal of the European Communities, L110A, 36.*
23. *Emergency Relief Items: Compendium of Basic Specifications: Medical Supplies and Equipment* (Vol. 2). UNDP/WHO, Geneva, 1996.
24. EU Council Directive 93/42/EEC of 14 June 1993 concerning medical devices. http://www.imdrf.org/docs/ghtf/archived/sg4/technical-docs/ghtf-sg4-guidelinesauditing-qms-part-1-general-requirements-080827.pdf (accessed on 29 May 2020).
25. European Commission, (2006). *Regulation (EC) No 1907/2006 of the European Parliament and of the Council of 18 December 2006 Concerning the Registration, Evaluation, Authorization and Restriction of Chemicals (REACH).* From http://eur-lex.europa.eu/legal-content/EN/TXT/?uri=CELEX%3A02006R1907-20140410 (accessed on 29 May 2020).
26. FDA, (2013). *Use of International Standards ISO-10993: Biological Evaluation of 3 Medical Devices Part 1: Evaluation and Testing.* Center for Devices and Radiological Health (CDRH). http://www.fda.gov/downloads/medicaldevices/deviceregulationandguidance/guidancedocuments/ucm348890.pdf (accessed on 29 May 2020).
27. Global Medical Devices Nomenclature System (GMDN) [website] https://www.gmdnagency.org (accessed on 29 May 2020).
28. Good governance for medicines. Model framework, (2014) Geneva: World Health Organization. http://apps.who.int/iris/bitstream/10665/129495/1/9789241507516_eng.pdf?ua=1&ua=1 (accessed on 29 May 2020).
29. Guiding principles for regulatory quality and performance, (2005). Paris: Organization for Economic Cooperation and Development (OECD). http://www.oecd.org/fr/reformereg/34976533.pdf (accessed on 29 May 2020).
30. Heimann, P., & Poluta, M. A., (1997). *Health Technology Management in the Sub-Saharan Region as a Prerequisite for Optimizing the Donor Aid Intervention Process.* (In press) WHO, ARA, Geneva.
31. ISO 10993-1, (2009). *Biological Evaluation of Medical Devices—*

Part 1: Evaluation and Testing Within a Risk Management Process. International Organization for Standardization. Geneva, Switzerland.

32. ISO 10993-10. (2010) *Biological Evaluation of Medical Devices—Part 10: Tests for Irritation and Skin Sensitization.* International Organization for Standardization. Geneva, Switzerland.
33. ISO 10993-11, (2017). *Biological Evaluation of Medical Devices—Part 11: Tests for Systemic Toxicity.* International Organization for Standardization. Geneva, Switzerland.
34. ISO 10993-12. (2012) *Biological Evaluation of Medical Devices—Part 12: Sample Preparation and Reference Materials.* International Organization for Standardization. Geneva, Switzerland.
35. ISO 10993-13. (2010) *Biological Evaluation of Medical Devices—Part 13: Identification and Quantification of Degradation Products from Polymeric Medical Devices.* International Organization for Standardization. Geneva, Switzerland.
36. ISO 10993-14. (2001) *Biological Evaluation of Medical Devices—Part 14: Identification and Quantification of Degradation Products from Ceramics.* International Organization for Standardization. Geneva, Switzerland.
37. ISO 10993-15. (2019) *Biological Evaluation of Medical Devices—Part 15: Identification and Quantification of Degradation Products from Metals and Alloys.* International Organization for Standardization. Geneva, Switzerland.
38. ISO 10993-16. (2017) *Biological Evaluation of Medical Devices—Part 16: Toxicokinetic Study Design for degradation products and Leachable.* International Organization for Standardization. Geneva, Switzerland.
39. ISO 10993-17. (2002) *Biological Evaluation of Medical Devices—Part 17: Establishment of Allowable Limits for Leachable Substances.* International Organization for Standardization. Geneva, Switzerland.
40. ISO 10993-18. (2005) *Biological Evaluation of Medical Devices—Part 18: Chemical Characterization of Materials.* International Organization for Standardization. Geneva, Switzerland.
41. ISO 10993-3. (2014) *Biological Evaluation of Medical Devices—Part 3: TESTS for Genotoxicity, Carcinogenicity and Reproductive Toxicity.* International Organization for Standardization. Geneva, Switzerland.
42. ISO 10993-4. (2017) *Biological Evaluation of Medical Devices—*

Part 4: Selection of Tests for Interactions with Blood. International Organization for Standardization. Geneva, Switzerland.

43. ISO 10993-5. (2009) *Biological Evaluation of Medical Devices—Part 5: Tests for In Vitro Cytotoxicity.* International Organization for Standardization. Geneva, Switzerland.
44. ISO 10993-6. (2016) *Biological Evaluation of Medical Devices—Part 6: Tests for Local Effects After Implantation.* International Organization for Standardization. Geneva, Switzerland.
45. ISO 10993-7. (2008) *Biological Evaluation of Medical Devices—Part 7: Ethylene Oxide Sterilization Residuals.* International Organization for Standardization. Geneva, Switzerland.
46. ISO 10993-9. (2019) *Biological Evaluation of Medical Devices—Part 9: Framework for Identification And Quantification of Potential Degradation Products. International Organization for Standardization.* Geneva, Switzerland.
47. ISO 13485. (2016) *Quality Management for Medical Devices.* International Organization for Standardization, Geneva, Switzerland. https://www.iso.org/obp/ui/#iso:pub:PUB100377 (accessed on 29 May 2020).
48. ISO 14155. (2011) *Clinical Investigation of Medical Devices for Human Subjects-Good Clinical Practice.* International Organization for Standardization, Geneva, Switzerland.
49. ISO 14971. (2019) *Application of Risk Management to Medical Devices.* International Organization for Standardization, Geneva, Switzerland.
50. ISO/TS 10993-20. (2006) *Biological evaluation of Medical Devices—Part 20: Principles and Methods Forimmunotoxicology Testing of Medical Devices.* International Organization for Standardization. Geneva, Switzerland.
51. Kramer, D., Xu, S., & Kessleheim, A., (2012). Regulation of medical devices in the United States and the European Union. *New Engl. J Med., 366*, 848–855. doi: 10.1056/NEJMhle1113918.
52. Magnusson, B., & Kligman, A. M., (1970). *Allergic Contact Dermatitis in the Guinea Pig* (pp. 102–123). Identification of contact allergens. Springfield, IL: Charles C Thomas.
53. Marzulli, F. N., & Maibach, H. I., (1975). The rabbit as a model for evaluating skin irritants: A comparison of results obtained on animals

and man using repeat skin exposures. *Food Cosmet. Toxicol., 13*, 533–540. doi: 10.1016/S0015-6264(78)80051-0.

54. McGarry, H. F., (2007). "The murine local lymph node assay: regulatory and potency considerations under REACH." *Toxicology, 238*(2, 3), 71–89.
55. Medical device regulations: global overview and guiding principles, (2003). Geneva: World Health Organization. http://www.who.int/medical_devices/publications/en/MD_Regulations.pdf (accessed on 29 May 2020).
56. Nancy, J. S., (1994). *Biocompatibility Testing and Management*. Clinical Design Group, Chicago.
57. NIH, (1999). *The Murine Local Lymph Node Assay: a Test Method for Assessing the Allergic Contact Dermatitis Potential of Chemicals and Compounds (publication no. 99-4494)*. Internal NIH publication.
58. Northup, S. J., (1999). Safety evaluation of medical devices: US Food and Drug Administration and International Standards Organization Guidelines. *International Journal of Toxicology, 18*, 275–283.
59. OECD Guidelines for testing of chemicals. (2018) *Part 451: Carcinogenicity Studies*. Geneva, Switzerland: Organization for Economic Cooperation and Development.
60. OECD Guidelines for testing of chemicals. (2018) *Part 453: Combined Chronic Toxicity/Carcinogenicity Studies*. Geneva, Switzerland: Organization for Economic Cooperation and Development.
61. OECD Guidelines for testing of chemicals. (1997) *Part 471: Genetic toxicology: Salmonella Typhimurium, Reverse Mutation Assay*. Geneva, Switzerland: Organization for Economic Cooperation and Development.
62. OECD Guidelines for testing of chemicals. (2014) *Part 473: Genetic Toxicology: In Vitro Mammalian Cytogenetic Test*. Geneva, Switzerland: Organization for Economic Cooperation and Development.
63. OECD Guidelines for testing of chemicals. (2018) *Part 479: Genetic Toxicology: In Vitro Sister Chromatid Exchange Assay in Mammalian Cells*. Geneva, Switzerland: Organization for Economic Cooperation and Development.
64. OECD, (1997). *OECD Guideline for the Testing of Chemicals: In Vitro Mammalian Cell Gene Mutation Test*. Available online at: http://www.oecd-ilibrary.org/docserver/download/9747601e.pdf?expires=142

(accessed on 29 May 2020).

65. Principles of conformity assessment for medical devices, (2012). Global Harmonization Task Force. GHTF/SG1/N78:2012; http://www.imdrf.org/docs/ghtf/final/sg1/technical-docs/ghtf-sg1-n78-2012-conformity-assessment-medicaldevices-121102.pdf. (accessed on 29 May 2020).
66. Principles of medical devices classification (2012). *Global Harmonization Task Force*; GHTF/SG1/N77:2012; http://www.imdrf.org/docs/ghtf/final/sg1/technicaldocs/ghtf-sg1-n77-2012-principles-medical-devices-classification-121102.pdf (accessed on 29 May 2020).
67. Quality management system—–Medical Devices—Guidance on corrective action and preventive action and related QMS processes (2010). *Global Harmonization Task Force* GHTF/SG3/N18:2010; http://www.imdrf.org/docs/ghtf/final/sg3/technical-docs/ghtf-sg3-n18-2010-qms-guidance-on-corrective-preventativeaction-101104.pdf (accessed on 29 May 2020).
68. Regulatory framework for control of refurbished medical devices (2012). Asia Pacific Economic Cooperation. http://mddb.apec.org/Documents/2012/MAG/WKSP1/12_mag_wksp1_011.pdf (accessed on 29 May 2020).
69. Regulatory harmonization and convergence. Silver Spring (MD): US Food and Drug Administration (US FDA) http://www.fda.gov/BiologicsBloodVaccines/InternationalActivities/ucm271079.htm (accessed on 29 May 2020).
70. Ritz, H. L., & Buehler, E. V., (1980). Planning, conduct and interpretation of guinea pig sensitization patch tests. In: Drill, V. A., & Lazar, P., (eds.), *Current Concepts in Cutaneous Toxocity* (pp. 25–40). New York: Academic Press.
71. Robinson, M. K., Nusair, T. L., Fletcher, E. R., & Ritz, H. L., (1990). A review of the Buehler guinea pig skin sensitization test and its use in a risk. assessment process for human skin sensitization. *Toxicology, 61*, 91–107.
72. Role of standards in the assessment of medical devices, (2008). *Global Harmonization Task Force*; 2008 GHTF/SG1/N044:2008; http://www.imdrf.org/docs/ghtf/final/ sg1/procedural-docs/ghtf-sg1-n044-2008-standards-in-assessment-of-medicaldevices-080305.pdf (accessed on 29 May 2020).

73. Rosenblatt, W. H., (1997). *Surplus Medical Supplies*. To be published in World Health Magazine.
74. Safety aspects—Guidelines for their inclusion in standards (2014). Geneva: International Organization for Standardization; ISO/IEC Guide 51:2014; http://www.iso.org/iso/catalogue_detail.htm?csnumber=53940 (accessed on 29 May 2020).
75. Shayne, C. G., (2002). *Safety Evaluation of Medical Devices*. Marcel Dekker, Inc., New York.
76. US EPA, (1998a). *Health Effects Test Guidelines OPPTS 870.1200 Acute Dermal Toxicity*. EPA 712-C-98-192. from https://www.regulations.gov/document?D=EPA-HQ-OPPT-2009-0156-0005 (accessed on 29 May 2020).
77. US EPA, (2002a). *Health Effects Test Guidelines OPPTS 870.1000 Acute Oral Toxicity Testing-Background*. EPA 712–C–02-189. from https://www.regulations.gov/document?D=EPA-HQOPPT-2009-0156-0002 (accessed on 29 May 2020).
78. US EPA. *Health Effects Test Guidelines OPPTS 870.1100 Acute Oral Toxicity*. EPA 712–C-02.
79. WHO global governance http://www.who.int/healthpromotion/conferences/9gchp/good-governance/en/ (accessed on 29 May 2020).
80. WHO good regulatory practices: guidelines for national regulatory authorities for medical products. Geneva: World Health Organization (under development, 2017).

INDEX